LOTTERIES

LOTTERIES

Prize Structures
and
Probabilities

Jaime Aguirre

Design and layout: Jaime Aguirre
Graphics: Jorge Flórez
ISBN 978-0-6453040-3-9

Independently published
Author's e-mail address: jaime.aguirre@unswalumni.com
Author's website: jaimeaguirre.au

CONTENTS

6

DRAWS WITHOUT REPLACEMENT: TWO DRAW MACHINES 181

7

SELECTED TOPICS — 243

BIBLIOGRAPHY — 311

INTRODUCTION

When you select to play fifteen numbers in a keno game, the chance of matching fifteen out of the twenty numbers drawn is 1 in 428,010,179,098. That is 1 in 428 billion! The number in itself is hard to imagine. Keno is a type of lottery, and as such, it is ruled by the laws of probability.

Lotteries are played in many countries around the world. Millions of people buy tickets on a regular or occasional basis in hopes of beating the odds. But how many of those dreamers have a real understanding of the mathematical principles ruling their chances? In the following pages you will find the basic information needed to clearly understand chance in terms of easy mathematical rules, and you will learn how to apply this knowledge to work out the chance of winning a prize in any lottery.

This book is addressed not only to students with a passion for the mathematics of chance, but also to those who simply want to understand how lotteries work. It is not an in-depth study of the subject, but rather an introduction to basic concepts and methods in probability, and how to apply them to most of the lotteries currently played worldwide.

We are constantly being persuaded to gamble through lottery websites, online games, and advertising. Understanding the risks of betting, and preventing the compulsion to gamble, begins with education. Education on wagering must start at the core of the matter: the knowledge of probability theory and the relevant mathematical rules applied to evaluate the chances of winning any lottery prize.

This book does not promise you any secret or system to win the lottery. If there were a systematic way of predicting a winning combination, the numerous authors of the large number of books on lottery systems would now be millionaires themselves, and would not be revealing their concealed secrets at an affordable value.

The subjects of the different chapters are:

Chapter 1: Lotteries, Odds, and Probabilities

The book starts with the basic concepts required to understand the laws of chance.

Chapter 2: Probability of Simple and Compound Events

In the second chapter the probability of simple and compound events is explained through several examples. The two fundamental principles of addition and multiplication are described.

Chapter 3: Methods of Counting

This chapter examines the permutations and combinations as techniques of counting. Counting methods are frequently used when calculating large quantities of outcomes.

Chapter 4: Draws with Replacement

Based on the definitions studied in Chapter 1 lotteries are classified in accordance with the probability principle involved. Draws with repetition are the equivalent to draws with replacement. The concept is explained with various examples.

Chapter 5: Draws without Replacement—One Draw Machine

Following the line of reasoning in the former chapter, lotteries in which repetitions do not occur are studied as draws without replacement, and in this chapter, lotteries using one draw machine are studied as examples.

Chapter 6: Draws without Replacement—Two Draw Machines

Draws without repetition using two draw machines are considered in this chapter. Again, all calculations are explained by way of several examples.

Chapter 7: Selected Topics

Finally, certain interesting issues are exposed for the reader to reflect on some aspects of lotteries that perhaps never crossed his/her mind before.

CHAPTER 1

LOTTERIES, ODDS, AND PROBABILITIES

"O Fortuna,
velut luna
statu variabilis,
semper crescis
aut decrescis…"

"Oh Fortune,
like the moon,
you are changeable,
ever waxing
and waning…"

Carmina Burana
(Profane Songs)

Lotteries

A lottery is a form of gambling based on chance, rather than skill, in which participants pay money for a chance to win one or more prizes. The distribution of prizes is determined by a predefined set of rules outlined in the prize structure. Typically, lotteries offer a range of cash prizes that correspond to different combinations or permutations of the winning numbers. State governments and organizations commonly sponsor lotteries as a means of generating funds. The methods for determining the winning numbers have many variations, but essentially most of them involve matching a combination of numbers drawn by a ball drawing machine, or a number randomly generated by a computer (a random number generator).

Lottery draws are essentially random experiments, characterized by a well-defined set of rules, where the outcomes are entirely dependent on chance and cannot be accurately predicted. Similar games of chance include slot machines (known as poker machines in Australia), instant lotteries (commonly referred to as scratch-off or scratch-and-win), and Keno. Since all these games rely on randomness, they follow the principles outlined in the mathematical theory of probability.

Chance

Uncertainty, or chance, is an inherent part of our daily lives. When we venture out of our homes in the early morning, we can never be certain of the events that may unfold. Surprisingly, even when we choose to stay at home, unexpected occurrences can still arise!

Chance represents the possibility or probability of various outcomes. It embodies the concept of uncertainty, standing in contrast to certainty. The study of chance falls within the realm of probability theory, a branch of mathematics that offers mathematical tools to understand what one can anticipate when events depend on possibilities.

Having a solid understanding of probability principles can be advantageous when participating in lotteries. Not because these principles can predict the winning numbers, but because they provide guidance on how to approach the game wisely and methodically. By applying probability principles, you can make informed decisions and enhance your overall lottery-playing strategy.

Probability

Probability is the branch of mathematics dealing with how likely something is to happen. In other words, probability studies mathematical procedures to compute chance.

Probability deals with the likelihood of a given event. The chance that an event might happen is numerically expressed as a fraction and is defined as follows: The probability of an outcome to an event E, provided that all outcomes are equally likely, is the number of cases favorable to that outcome divided by the total number of cases possible.

$$\text{Probability of event } E = \frac{Number\ of\ favorable\ outcomes}{Total\ number\ of\ possible\ outcomes}$$

This definition is condensed in the formula

$$P(E) = \frac{n}{N}$$

$P(E)$ is the probability of the event E,
n is the number of favorable outcomes, and
N is the total number of possible outcomes.

EXAMPLE 1.1

A box contains three red marbles and two white marbles. When drawing one marble at random, what is the likelihood of obtaining a white marble?

SOLUTION

n = 2 (two white marbles)

N = 5 (two white marbles plus three red marbles)

$$P(E) = \underline{\hspace{8cm}}$$

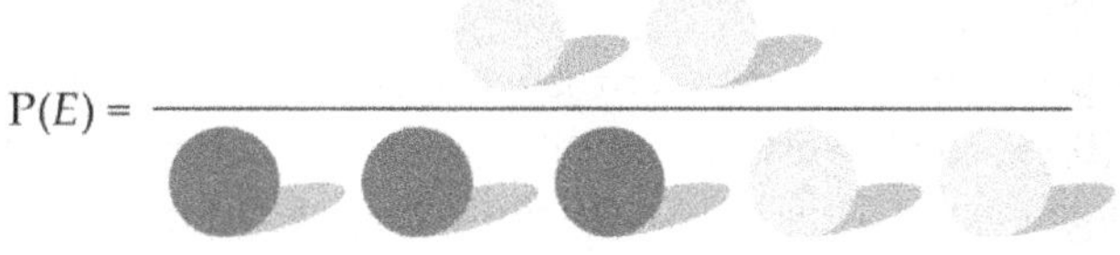

$$P(E) = \frac{n}{N} = \frac{2}{5} = 0.4 = 40\%$$

EXAMPLE 1.2

What is the probability of obtaining heads when tossing a coin?

SOLUTION

n (number of favorable outcomes) = 1 (one head side)

N (number of possible outcomes) = 2 (two sides)

$$P(E) = \underline{\hspace{8cm}}$$

$$P(E) = \frac{n}{N} = \frac{1}{2} = 0.5 = 50\%$$

Other examples of events are tossing coins, throwing dice, drawing cards from a standard deck of cards, and drawing marbles from a barrel.

As all outcomes are equally likely, the experiments performed must not be biased: coins must be fair (not bent or weighted) and tossed fairly, dice must be true, decks of cards must be shuffled, and any draw of a card or marble must be done at random.

The Probability Scale

The probability of an event is expressed as a real number within the interval [0,1]. The lowest probability is zero, which represents an event that can never occur. This is the absolute impossibility of an event. An example of this is the probability of you winning a prize (event A) in a particular lottery if you did not buy a ticket. It is not possible for you to win any prize.

$$P(A) = \frac{0}{N} = 0$$

P(*winning a prize*) = 0

The highest probability is one, which represents an event that will certainly occur. This is the absolute certainty of an event. For example, if you bought so many tickets that you had in your possession the total number of possible combinations in a particular lottery, the probability that you will have the winning combination (event B) is one. You will certainly win.

$$P(B) = \frac{N}{N} = 1$$

P(*winning the jackpot*) = 1

It becomes axiomatic that any probability value must occur between zero and one. The lower the probability, the closer it is to zero. The higher the probability, the closer it is to one. In other words, events that are highly unlikely to occur have a probability close to zero, and events that are highly likely to occur have a probability close to one.

The probability scale is represented in figure 1.1 as an analogy to the mythical Tower of Babel: the probability that people in ancient times were able to build a tower reaching the troposphere (the lowest portion of Earth's atmosphere) was zero.

Figure 1.1 The Tower of Probability

Probability is usually expressed as a fraction or a decimal, but it can be converted into a percentage. Thus, a probability of 2/5 is equivalent to 0.4 or 40 percent.

EXAMPLE 1.3

What is the probability of throwing the number six when rolling a six-sided die?

SOLUTION

n (number of favorable outcomes) = 1 (one side)

N (number of possible outcomes) = 6 (six sides)

$$P(E) = \frac{n}{N} = \frac{1}{6} = 0.167 = 16.7\%$$

EXAMPLE 1.4

What is the probability of drawing any ace when selecting one card at random from a full deck of cards?

SOLUTION

n (number of favorable outcomes) = 4 (four aces)

N (number of possible outcomes) = 52 (fifty-two cards)

$$P(E) = \frac{n}{N} = \frac{4}{52} = \frac{1}{13} = 0.077 = 7.7\%$$

Odds

In games of chance probabilities are also expressed in terms of odds. The odds of an event occurring is the ratio of the number of ways the event can occur (successes) to the number of ways the event cannot occur (failures). The odds of an event can also be defined as the ratio of the probability of the event happening to the probability of the event not happening.

Odds in Favor

The odds in favor of the event E are expressed as the ratio of the number of favorable outcomes to the number of unfavorable outcomes. This is written as

$$O_{for}(E) = \frac{n}{(N-n)}$$

$O_{for}(E)$ = the odds in favor of event E,
N = total possible outcomes, and
n = favorable outcomes.

Odds Against

The odds against the event E are expressed as the ratio of the number of unfavorable outcomes to the number of favorable outcomes. This is written as

$$O_{against}(E) = \frac{(N-n)}{n}$$

$O_{against}(E)$ = the odds against event E,
N = total possible outcomes, and
n = favorable outcomes.

As it might be expected, the odds for the event happening are the reciprocal of those against. Thus, for instance, if event E has odds 10 to 1 against, it has odds 1 to 10 in favor.

$$O_{for}(E) \times O_{against}(E) = 1$$

<u>EXAMPLE 1.5</u>

When tossing one coin, what are the odds of obtaining heads?

<u>SOLUTION</u>

n (number of favorable outcomes) = 1

N (number of possible outcomes) = 2

$$O_{for}(\text{heads}) = \frac{n}{(N-n)} = \frac{1}{(2-1)} = \frac{1}{1}$$

Also written as 1:1 (read as "1 to 1")

$$O_{against}(\text{heads}) = \frac{(N-n)}{n} = \frac{(2-1)}{1} = \frac{1}{1}$$

Also written as 1:1 (read as "1 to 1")

<u>EXAMPLE 1.6</u>

When rolling one die (six-sided), what are the odds of obtaining the number four?

<u>SOLUTION</u>

n (number of favorable outcomes) = 1

N (number of possible outcomes) = 6

$$O_{for}(\textit{number four}) = \frac{n}{(N-n)} = \frac{1}{(6-1)} = \frac{1}{5}$$

Also written as 1:5 (read as "1 to 5")

$$O_{against}(\textit{number four}) = \frac{(N-n)}{n} = \frac{(6-1)}{1} = \frac{5}{1}$$

Also written as 5:1 (read as "5 to 1")

<u>EXAMPLE 1.7</u>

One card is drawn at random from a full deck of cards. What are the odds of choosing an ace?

<u>SOLUTION</u>

n (number of favorable outcomes) = 4

N (number of possible outcomes) = 52

$$O_{for}(ace) = \frac{n}{(N-n)} = \frac{4}{(52-4)} = \frac{4}{48} = \frac{1}{12}$$

Also written as 1:12 (read as "1 to 12")

$$O_{against}(ace) = \frac{(N-n)}{n} = \frac{(52-4)}{4} = \frac{48}{4} = \frac{12}{1}$$

Also written as 12:1 (read as "12 to 1")

Difference Between Odds and Probability

$$Probability\ of\ event\ E = \frac{Number\ of\ favorable\ outcomes}{Total\ number\ of\ possible\ outcomes}$$

Remember that the total number of possible outcomes is the sum of the favorable outcomes plus the unfavorable outcomes.

$$Odds\ in\ favor\ of\ event\ E = \frac{Number\ of\ favorable\ outcomes}{Number\ of\ unfavorable\ outcomes}$$

It is a common practice in lottery websites to denote odds as

$$1\ in\ \frac{Total\ number\ of\ possible\ outcomes}{Number\ of\ favorable\ outcomes}$$

This expression is actually the probability. For a more detailed explanation see:

Fulton, L V et al. Confusion between odds and probability, a pandemic? *Journal of Statistics Education*, volume 20, number 3, 2012.

JAIME AGUIRRE

CHAPTER 2

PROBABILITY OF SIMPLE AND COMPOUND EVENTS

Simple Events

Tossing a coin once, rolling a die once, or drawing one card from a full deck are single events. Only one event occurs.

The probability of single events is calculated by applying the definition formula:

$$P(E) = \frac{n}{N}$$

EXAMPLE 2.1

When tossing one coin, what is the probability of obtaining tails?

SOLUTION

n (number of favorable outcomes) = 1 (one tail side)

N (number of possible outcomes) = 2 (two sides)

$$P(E) = \frac{n}{N} = \frac{1}{2} = 0.50 = 50\%$$

EXAMPLE 2.2

When rolling one die (six-sided), what is the probability of obtaining the number one?

SOLUTION

n (number of favorable outcomes) = 1 (one side)

N (number of possible outcomes) = 6 (six sides)

$$P(E) = \frac{n}{N} = \frac{1}{6} = 0.167 = 16.7\%$$

EXAMPLE 2.3

When drawing one marble from a box containing two red marbles and two white marbles, what is the probability of drawing a white marble?

SOLUTION

n (number of favorable outcomes) = 2 (two white marbles)

N (number of possible outcomes) = 4 (four total marbles)

$$P(E) = \frac{n}{N} = \frac{2}{4} = \frac{1}{2} = 0.5 = 50\%$$

EXAMPLE 2.4

One card is drawn at random from the suit of clubs. What is the probability that the card chosen will be the ace of clubs?

SOLUTION

n (number of favorable outcomes) = 1 (the ace of clubs)

N (number of possible outcomes) = 13 (thirteen cards in a suit)

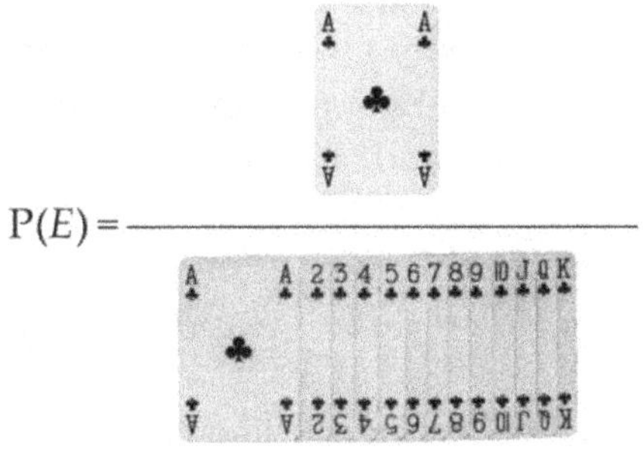

$$P(E) = \frac{n}{N} = \frac{1}{13} = 0.077 = 7.7\%$$

EXAMPLE 2.5

The numbers one through five are each written on one of five cards, and the cards are thoroughly mixed in a bag. One card is drawn at random. What is the probability of drawing the card with the number three on it?

SOLUTION

n (number of favorable outcomes) = 1 (one card with the number 3)

N (number of possible outcomes) = 5 (five cards)

$$P(E) = \frac{n}{N} = \frac{1}{5} = 0.20 = 20\%$$

Compound Events

The examples given in the previous section involve the probabilities when considering a single event. But it might happen that two or more events are to be considered when occurring either simultaneously or one after the other. Then we have compound events, and the addition and multiplication principles are applied to work out the probabilities of such events.

Two main types of compound events are commonly taken from ordinary events: considering two or more different outcomes when performing one experiment and considering two or more different outcomes when performing two or more experiments.

Examples of considering two or more different outcomes when performing one experiment include the probabilities of obtaining the following:

- ✓ heads or tails when tossing a coin once,
- ✓ the number one or the number two when throwing a die once,
- ✓ the ace of hearts or the ace of clubs when drawing a card from a suit, and
- ✓ a green marble or a blue marble when drawing a marble from a box.

Examples of situations considering two or more different outcomes when performing two or more experiments, include the following:

- ✓ tossing a coin twice or more,
- ✓ tossing two or more separate coins simultaneously or one after the other,
- ✓ rolling a die twice or more,
- ✓ rolling two or more separate dice simultaneously or one after the other,
- ✓ drawing two or more cards from a group of cards, and
- ✓ drawing two or more marbles from a box.

The Addition Principle

Mutually Exclusive Events

A box contains three red marbles and one black marble. When drawing one marble at random, what is the probability that it will be either red or black?

Because it has been stated that the marbles are either red or black, one marble cannot possibly be red and black at the same time. Therefore, when drawing one marble, the occurrence of one event (drawing a red marble) excludes the possibility of the other (drawing a black marble) happening, and vice versa.

These are mutually exclusive events.

Two events are mutually exclusive if they cannot occur simultaneously

---

The probability of such events is worked out by means of the addition principle, which states that the probability that one or another of several mutually exclusive events happens is the sum of the probabilities of the individual events.

If we have the events A and B, then

$$P(A \text{ or } B) = P(A) + P(B)$$

<u>EXAMPLE 2.6</u>

When tossing one coin, what is the probability of throwing either heads or tails?

<u>SOLUTION</u>

A = Heads is obtained.

B = Tails is obtained.

E = Either heads or tails is obtained.

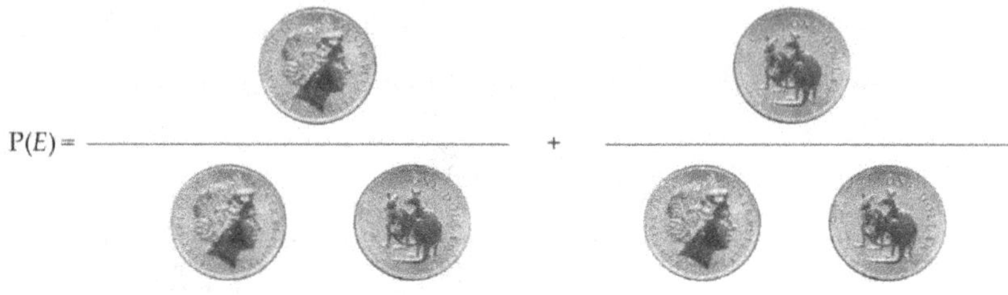

$$P(E) = P(A) + P(B) = \frac{1}{2} + \frac{1}{2} = \frac{2}{2} = 1 = 100\%$$

In other words, throwing either heads or tails is a certain event.

<u>EXAMPLE 2.7</u>

When rolling one die (six-sided), what is the probability of getting either one or two?

<u>SOLUTION</u>

A = The number one is obtained.

B = The number two is obtained.

E = Either the number one or the number two is obtained.

$$P(E) = P(A) + P(B) = \frac{1}{6} + \frac{1}{6} = \frac{2}{6} = \frac{1}{3} = 0.33 = 33\%$$

EXAMPLE 2.8

One card is drawn at random from a set of cards comprising the four aces. What is the probability that the card chosen will be either the ace of hearts or the ace of diamonds?

SOLUTION

A = The ace of hearts is obtained.

B = The ace of diamonds is obtained.

E = Either the ace of hearts or the ace of diamonds is obtained.

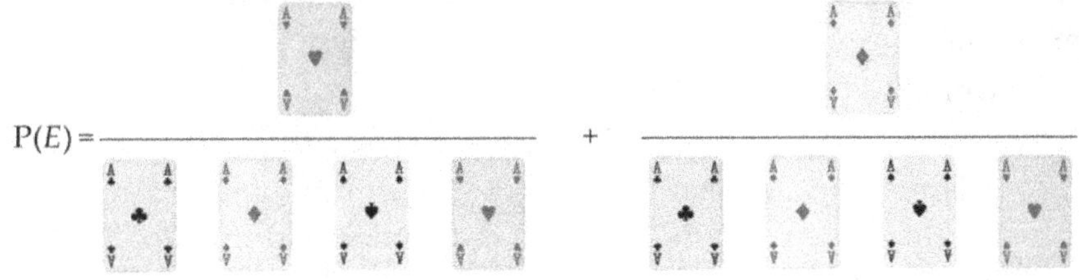

$$P(E) = P(A) + P(B) = \frac{1}{4} + \frac{1}{4} = \frac{2}{4} = \frac{1}{2} = 0.50 = 50\%$$

EXAMPLE 2.9

One marble is drawn at random from a box containing three green marbles and one blue marble. What is the probability that it will be either green or blue?

SOLUTION

A = A green marble is obtained.

B = A blue marble is obtained.

E = Either a green marble or a blue marble is obtained.

$$P(E) = P(A) + P(B) = \frac{3}{4} + \frac{1}{4} = \frac{4}{4} = 1 = 100\%$$

The Multiplication Principle

Case A: Independent Events

<u>Draws with Replacement</u>

A box contains two white marbles and three red marbles. One marble is drawn and then placed back in the box; then another is drawn. Find the probabilities of obtaining two white marbles.

Because there are two favorable outcomes (two white marbles) and five possible outcomes (five marbles), the probability that the first marble will be white is P(*white marble*) $= \dfrac{2}{5}$

Because the first marble is replaced before drawing the second one, the two events have no influence upon each other, and the probability of the second marble being white is P(*white marble*) $= \dfrac{2}{5}$

It is important to note that this case is equivalent to drawing one marble from one box containing two white marbles and three red marbles, and then drawing one marble from another box also containing two white marbles and three red marbles.

These are independent events.

> *Two events are independent if the occurrence of one*
> *does not affect the occurrence of the other*

The probability of independents events is worked out by means of the multiplication principle, which states that the probability that two independent events will happen is the product of the probabilities of the individual events.

If we have the events A and B, then

$$P(A \text{ and } B) = P(A) \times P(B)$$

EXAMPLE 2.10

When tossing one coin twice, what is the probability of obtaining two heads?

SOLUTION

A = Heads is obtained in the first toss.

B = Heads is obtained in the second toss.

E = Heads is obtained in both tosses.

$$P(E) = P(A) \times P(B) = \frac{1}{2} \times \frac{1}{2} = \frac{1}{4} = 0.25 = 25\%$$

EXAMPLE 2.11

When rolling one die twice, what is the probability of throwing a three and then a four?

SOLUTION

A = The number three is obtained in the first throw.

B = The number four is obtained in the second throw.

E = The number three is obtained in the first throw, and the number four is obtained in the second throw.

$$P(E) = P(A) \times P(B) = \frac{1}{6} \times \frac{1}{6} = \frac{1}{36} = 0.028 = 2.8\%$$

EXAMPLE 2.12

One card is drawn from a set of cards comprising the four aces. It is replaced and then another one is drawn. What is the probability of obtaining the ace of hearts and then the ace of clubs?

SOLUTION

A = The ace of hearts is obtained in the first draw.

B = The ace of clubs is obtained in the second draw.

E = The ace of hearts is obtained in the first draw, and the ace of clubs is obtained in the second draw.

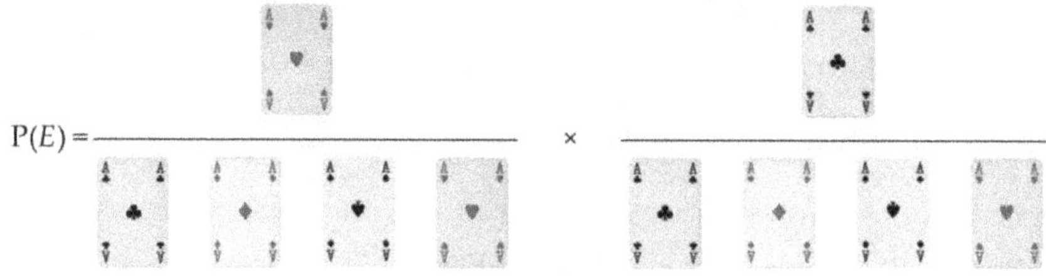

$$P(E) = P(A) \times P(B) = \frac{1}{4} \times \frac{1}{4} = \frac{1}{16} = 0.0625 = 6.25\%$$

EXAMPLE 2.13

A box contains two white marbles and three red marbles. One marble is drawn and replaced, and then another one is drawn. Find the probability of obtaining two white marbles.

SOLUTION

A = A white marble is obtained in the first draw.

B = A white marble is obtained in the second draw.

E = A white marble is obtained in the first draw and in the second draw.

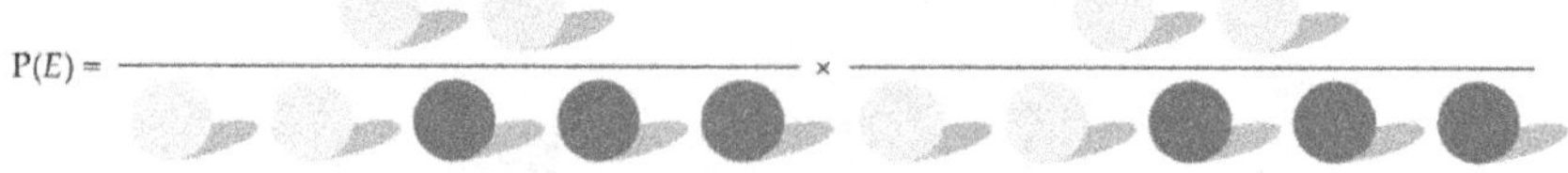

$$P(E) = P(A) \times P(B) = \frac{2}{5} \times \frac{2}{5} = \frac{4}{25} = 0.16 = 16\%$$

EXAMPLE 2.14

Suppose there are two barrels, each containing five balls numbered from 1 to 5. When drawing one ball from each barrel, what is the probability of obtaining the combination 4 and 5?

SOLUTION

A = The number 4 is obtained in the first draw.

B = The number 5 is obtained in the second draw.

E = The number 4 is obtained in the first draw, and the number 5 is obtained in the second draw.

$$P(E) = \frac{4}{1 \quad 2 \quad 3 \quad 4 \quad 5} \times \frac{5}{1 \quad 2 \quad 3 \quad 4 \quad 5}$$

$$P(E) = P(A) \times P(B) = \frac{1}{5} \times \frac{1}{5} = \frac{1}{25} = 0.04 = 4\%$$

Case B: Dependent Events

<u>Draws without Replacement</u>

Consider again a box containing two white marbles and three red marbles. One marble is drawn but not replaced. Then another marble is drawn. What is the probability of getting two white marbles?

The probability of the first marble being white is

$$P(\textit{white marble}) = 2/5$$

Given the condition that the first draw has resulted in a white marble, the probability of the second marble being white is

$$P(\textit{white marble}) = 1/4$$

because there are now four marbles in the box, one of which is white. Remember that the first marble drawn was not replaced.

These are dependent events.

> *Two events are dependent if the occurrence of one*
> *affects the occurrence of the other.*

The probability of such events is also worked out by means of the multiplication principle as follows: the probability that two dependent events both happen is the product of the probabilities of the individual events, the latter being calculated on the assumption that the first event has already happened.

If we have the dependent events A and B, then

P(A then B) = P(A) × P(B, *given that A has occurred*)

EXAMPLE 2.15

Two cards are drawn without replacement from a set of cards comprising the four aces. What is the probability of obtaining the ace of hearts and then the ace of clubs?

SOLUTION

A = The ace of hearts is obtained in the first draw.

B = The ace of clubs is obtained in the second draw.

E = The ace of hearts is obtained in the first draw and the ace of clubs in the second.

$$P(E) = P(A) \times P(B, \text{ given that } A \text{ has occurred}) = \frac{1}{4} \times \frac{1}{3} = \frac{1}{12} = 0.083 = 8.3\%$$

EXAMPLE 2.16

A box contains two white marbles and three red marbles. Two marbles are drawn without replacement. What is the probability of obtaining two white marbles?

SOLUTION

A = A white marble is obtained in the first draw.

B = A white marble is obtained in the second draw.

E = A white marble is drawn in both draws.

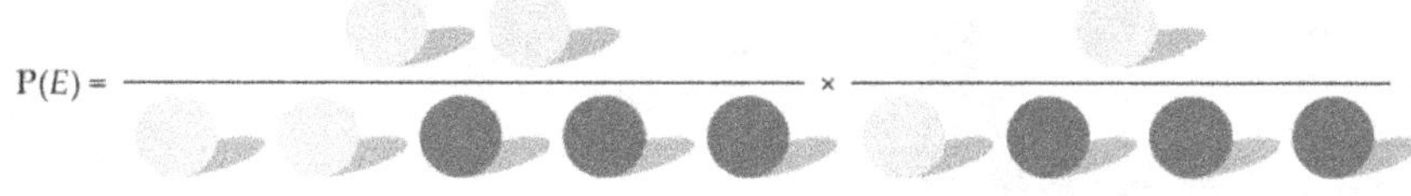

$$P(E) = P(A) \times P(B, \textit{given that A has occurred}) = \frac{2}{5} \times \frac{1}{4} = \frac{2}{20} = \frac{1}{10} = 0.10 = 10\%$$

EXAMPLE 2.17

One box contains six balls numbered from 1 to 6. Two balls are drawn without replacement. What is the probability of obtaining the number 6, and then the number 3?

SOLUTION

A = The number 6 is obtained in the first draw.

B = The number 3 is obtained in the second draw.

E = The number 6 is obtained in the first draw, and the number 3 is obtained in the second draw.

$$P(E) = P(A) \times P(B, \textit{given that A has occurred}) = \frac{1}{6} \times \frac{1}{5} = \frac{1}{30} = 0.033 = 3.3\%$$

CHAPTER 3

METHODS OF COUNTING

As seen before, to be able to calculate the probability of one or more events, it is necessary to know the number of favorable outcomes n, and the number of total possible outcomes N.

In the events of tossing one or more coins, throwing one or more dice, or drawing one or more cards, the counting is easy. But when dealing with more complicated situations, counting methods involving combinations and permutations must be applied.

Permutations and Combinations

Permutations and combinations are mathematical tools used for counting. It is often extremely complicated to count the number of favorable outcomes or the total number of possible outcomes. By using permutations and combinations, the counting process can be simplified.

Consider the following experiment: An urn contains three differently colored marbles—namely white, black, and red. In how many ways can we take a sample of two marbles?

Before attempting to solve this problem, two questions must be asked: Is order important? And is the first marble to be replaced before drawing the second?

Is order important?

If the order in which the marbles are drawn is important, we will get an ordered arrangement, also called a permutation. If the order in which the marbles are drawn is not important, we will get an unordered arrangement, also called a combination.

Is the first marble to be replaced before draw the second one?

If the first marble drawn is to be replaced before drawing the second one, we will deal with the problem as a draw with replacement. If the first marble drawn is to be set aside before drawing the next one, we will deal with the problem as a draw without replacement.

Permutations

A permutation, also called an ordered arrangement, is a selection of a subset of objects from a set with regard to order. In other words, a permutation is an arrangement of objects or events where the order is important.

Permutations with Repetition

Draws with Replacement

If selections are made with replacement, permutations with repetition are formed. When we have n things to choose from, we have n choices each time. When choosing r of them the total number of possible permutations is equal to r times n.

In other words, there are n possibilities for the first choice, and then there are n possibilities for the second choice, and so on, multiplying each time. This is expressed by the equation

$$P(n,r) = n^r$$

in which P stands for permutations, n stands for the total number of objects to choose from, and r stands for the number of objects in the arrangement. This formula is also written as

$$nPr = n^r$$

EXAMPLE 3.1

An urn contains three marbles numbered four, five, and six. By drawing two marbles with replacement, how many different permutations can be obtained?

SOLUTION

The first marble drawn can be either of three numbers. If it is marble 4, then the second one can be either 4 (because the marble drawn was replaced), 5, or 6, and three different permutations can occur:

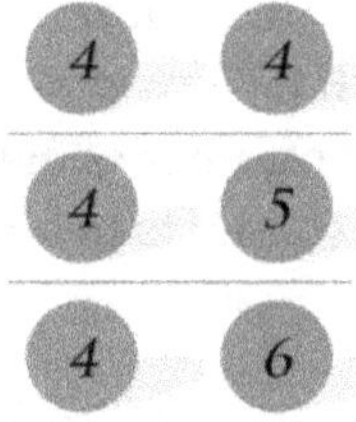

If the first marble drawn is marble 5, the second one can be either 5 (because the marble drawn was replaced), 4, or 6, and three more different permutations occur:

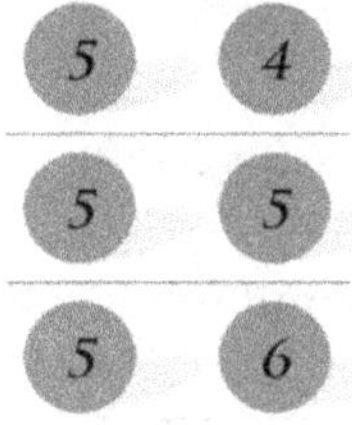

If the first marble drawn is marble 6, then the second one can be either 6 (because the marble drawn was replaced), 4, or 5, and three more permutations can occur:

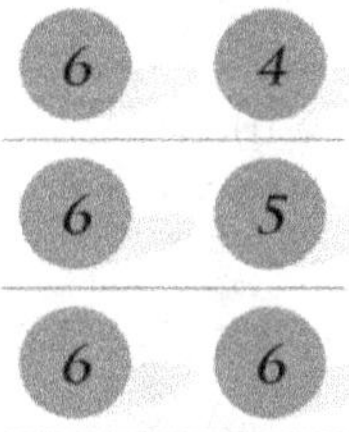

This has been a quite easy exercise, but what if we want to predict the number of different permutations when drawing six marbles with replacement, from an urn containing 44 marbles?

Although the reasoning will be always the same, the process of making the calculations would be tedious and time-consuming. A general formula is applied to calculate permutations with repetition:

$$P(n,r) = n^r$$

Where P(n,r) = the number of different permutations with repetition that
can be made with n elements, taken r at a time,
n = the total number of elements, and
r = the number of elements to be taken in each set.

In the above example,

$$P(3,2) = 3^2 = 9$$

In simple terms, to find out the permutations with repetition of r from n, multiply n by itself r times.

In the same example,

3 multiplied by itself two times: 3 x 3 = 9

And we have nine permutations with repetition for P(3,2).

EXAMPLE 3.2
An urn contains ten balls numbered one to ten. How many different permutations of numbers can be made when drawing six balls with replacement?

SOLUTION
r (number of balls drawn) = 6
n (total number of balls) = 10

$$P(10,6) = 10^6 = 1,000,000$$

10 multiplied by itself six times: 10 x 10 x 10 x 10 x 10 x 10 = 1,000,000

Permutations without Repetition

Draws without Replacement

If selections are made without replacement, permutations without repetition are formed.

EXAMPLE 3.3

An urn contains three marbles numbered four, five, and six. When drawing two marbles without replacement, how many permutations can we obtain?

SOLUTION

The first marble drawn can be either of three numbers. If it is marble 4, then the second can be either 5 or 6, and two different permutations can occur:

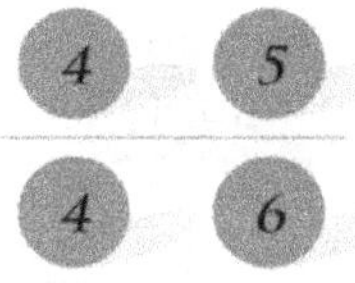

If the first marble drawn is marble 5, the second one can be either 4 or 6, and two more permutations can occur:

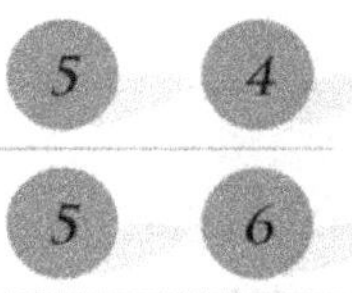

If the first marble drawn is marble 6, the second one can be either 4 or 5, and two more permutations can occur:

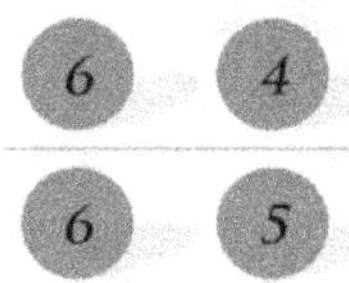

The formula to calculate permutations without repetition is

$$P(n,r) = \frac{n!}{(n-r)!}$$

P(n,r) = the number of permutations without repetition that can be made with n elements, taken r at a time,

r = set of elements taken at a time,

n = size of the total population.

The exclamation mark following a natural number is called a factorial symbol. The factorial of n, denoted by n! is the product of a sequence of natural numbers starting with n and ending with 1; n! is usually verbalized as "n factorial".

In the above example,

$$P(3,2) = \frac{3!}{(3-2)!} = \frac{3!}{1!} = \frac{3 \times 2 \times 1}{1} = 6$$

A simplified form of performing this calculation is used as follows: to find out the permutations without repetition of r from n, multiply n by every number smaller than itself until there are a number of factors equal to r.

In the same example: multiply 3 by every number smaller than itself until there are two factors:

$$3 \times 2 = 6$$

And we have six permutations without repetition for P(3,2).

EXAMPLE 3.4

An urn contains ten balls numbered one to ten. How many different permutations of numbers can be made when drawing six balls without replacement?

SOLUTION

r (number of balls drawn) = 6

n (total number of balls) = 10

$$P(10,6) = \frac{10!}{(10-6)!} = \frac{10 \times 9 \times 8 \times 7 \times 6 \times 5 \times 4 \times 3 \times 2 \times 1}{4 \times 3 \times 2 \times 1} = 151,200$$

Or multiply 10 by every number smaller than itself until you get a number of six factors:

$$P(10,6) = 10 \times 9 \times 8 \times 7 \times 6 \times 5 = 151,520$$

A Special Case of Permutations

A special case of permutations occurs when there are identical elements within the set. In this case the calculation is performed with the formula

$$P(n,r) = \frac{n!}{x! \times y! \times z!}$$

Where P(n,r) = the number of permutations without repetitions that can be made with n elements, taken r at a time,

r = set of elements taken at a time,

n = total number of elements,

x = number of identical elements A,

y = number of identical elements B,

z = number of identical elements C

EXAMPLE 3.5

How many different permutations can be made with the letters A A B B C?

SOLUTION

n = 5

r = 3

x = 2

y = 2

z = 1

$$P(5,3) = \frac{5!}{2! \times 2! \times 1!} = 30$$

Combinations

A combination, also called an unordered arrangement, is a selection of a subset of objects from a set without regard to order. In other words, a combination is an arrangement of objects or events where the order is not important.

Combinations with Repetition

<u>Draws with Replacement</u>

If selections are made with replacement, combinations with repetition are formed.

<u>EXAMPLE 3.6</u>

An urn contains three marbles numbered four, five, and six. When drawing two marbles with replacement, how many different combinations can we obtain?

<u>SOLUTION</u>

The first marble drawn can be any of three numbers. If it is marble 4, then the second one can be either 4 (because the white marble drawn was replaced), 5, or 6, and three different combinations can occur:

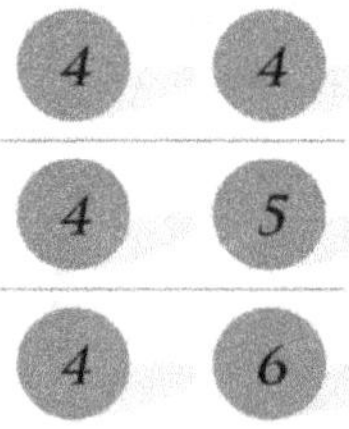

If the first marble drawn is marble 5, the second one can be either 5 (because the marble drawn was replaced), 4, or 6, and three combinations can occur. But because one of these combinations, marble 5 with marble 4, has already occurred (order is disregarded in this case) and has thus already been counted, only two new combinations are formed:

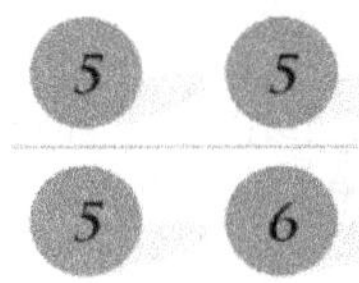

If the first marble drawn is 6, then the second one can be either 6 (because the marble drawn was replaced), 4, or 5, and three combinations can occur. But because two of these combinations, marble 5 with marble 6 and marble 4 with marble 6, have already occurred (order is disregarded in this case), and have thus already been counted, only one new combination is formed:

The formula to calculate combinations with repetition is

$$C(n,r) = \frac{(n+r-1)!}{r!(n-1)!}$$

$C(n,r)$ = number of different combinations with repetition that can be made with n elements, taken r at a time,
r = set of elements taken at a time,
n = total number of elements.

In the above example:

$$C(3,2) = \frac{(3+2-1)!}{2!(3-1)!} = \frac{4!}{2! \times 2!} = \frac{4 \times 3 \times 2 \times 1}{2 \times 1 \times 2 \times 1} = 6$$

EXAMPLE 3.7

An urn contains ten balls numbered one to ten. How many different combinations of numbers can be made when drawing six balls with replacement?

SOLUTION

r (number of balls drawn) = 6

n (total number of balls) = 10

$$C(10,6) = \frac{(10+6-1)!}{6!(10-1)!} = \frac{15!}{6! \times 9!} = \frac{15 \times 14 \times 13 \times 12 \times 11 \times 10 \times 9 \times 8 \times 7 \times 6 \times 5 \times 4 \times 3 \times 2 \times 1}{6 \times 5 \times 4 \times 3 \times 2 \times 1 \times 9 \times 8 \times 7 \times 6 \times 5 \times 4 \times 3 \times 2 \times 1} = 5{,}005$$

Combinations without Repetition

Draws without Replacement

If selections are made without replacement, combinations without repetition are formed.

EXAMPLE 3.8

An urn contains three marbles numbered four, five, and six. When draw two marbles without replacement, how many different combinations can be obtained?

SOLUTION

The first marble drawn can be any of three numbers. If it is marble 4, then the second one can be either 5 or 6, and two different combinations can occur:

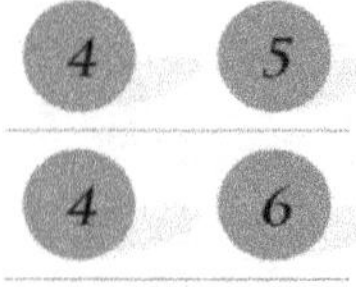

If the first marble drawn is marble 5, the second one can be either 4 or 6, and two combinations can occur. Because one of these combinations, marble 5 with marble 4, has already occurred (order is disregarded in this case), only one different combination is formed:

If the first marble drawn is marble 6, then the second one can be either 4 or 5, and two combinations can occur. Because both combinations have already occurred (order is disregarded), there are no new combinations, and the answer is that three different combinations without repetition can be made when drawing two marbles, without replacement, from an urn containing three differently colored marbles.

The formula to calculate combinations without repetition is

$$C(n,r) = \frac{n!}{r!(n-r)!}$$

$C(n,r)$ = the number of different combinations without repetition that can be made with n elements, taking r at a time.
r = set of elements taken at a time,
n = total number of elements.

In the above example,

$$C(3,2) = \frac{3!}{2!(3-2)!} = \frac{3 \times 2 \times 1}{2 \times 1 \times 1} = 3$$

There is a simplified form of performing this calculation: to find out the combinations without repetition of r from n, multiply r by every number smaller than itself. This becomes the denominator of the fraction. The numerator is n multiplied by each number smaller than itself until there are an equal number of factors in both the numerator and the denominator.

In the same example,
Numerator:
Multiply 3 by every number smaller than itself until there are two factors: 3 x 2

Denominator:
Multiply 2 by every number smaller than itself: 2 x 1

$$C(3,2) = \frac{3 \times 2}{2 \times 1} = 3$$

EXAMPLE 3.9

Consider that there are ten marbles numbered one to ten. How many different combinations of numbers can be made when drawing six marbles without replacement?

SOLUTION

r (number of balls drawn) = 6

n (total number of balls) = 10

$$C(10,6) = \frac{10!}{6!(10-6)!} = \frac{10\times9\times8\times7\times6\times5\times4\times3\times2\times1}{6\times5\times4\times3\times2\times1\times4\times3\times2\times1} = 210$$

Numerator:

Multiply 10 by every number smaller than itself until there are six factors: 10 x 9 x 8 x 7 x 6 x 5

Denominator:

Multiply 6 by every number smaller than itself: 6 x 5 x 4 x 3 x 2 x 1

$$C(10,6) = \frac{10\times9\times8\times7\times6\times5}{6\times5\times4\times3\times2\times1} = 210$$

The following different notations will be used in the subsequent chapters:

PERMUTATIONS	COMBINATIONS
$P_{(n,r)}$	$C_{(n,r)}$
$_nP_r$	$_nC_r$
P_r^n	C_r^n

Table 3.1 shows a comparison of the formulas applied and the results obtained in different examples on permutations and combinations.

TABLE 3.1 Summary of formulas to calculate
permutations and combinations

	Formula	Examples	
		$n=3$ $r=2$	$n=10$ $r=6$
Permutations with Repetition	$P(n,r) = n^r$	9	1,000,000
Permutations **without** Repetition	$P(n,r) = \dfrac{n!}{(n-r)!}$	6	151,200
Combinations with Repetition	$C(n,r) = \dfrac{(n+r-1)!}{r!(n-1)!}$	6	5,005
Combinations **without** Repetition	$C(n,r) = \dfrac{n!}{r!(n-r)!}$	3	210

CHAPTER 4

DRAWS WITH REPLACEMENT

Lotteries are compound events, in terms of probability, and they are usually either dependent or independent events. To calculate the probabilities of every division prize, we apply the formulas described in chapters 1, 2, and 3.

Draws with replacement include lotteries in which every number drawn is put back into the system before drawing the next number. Given the fact that this is not very practical, the actual procedure consists of drawing one number from each of several ball drawing machines, all containing the same numbers. This is exactly the same as drawing the numbers from one ball drawing machine and replacing the number before drawing the next one. So, they are actually draws with replacement, even if the replacement is not materially taking place. As it can be expected, this procedure may produce repeated numbers.

A lottery is not only defined by the probability event taking place, but also by the different rules established previously to the draw. In this manner, a range of division prizes is defined for every lottery. Even if the same probability events take place, the division prizes can be different, if distinct rules have been previously established. Separate rules can make two distinctive lotteries from the same probability event.

Every outcome can be defined in either of these two ways:

1. As a prize division of one lottery
2. As a lottery in itself

In each case the calculation of probabilities is different. What are the criteria to select either way? The betting rules defined prior to the draw. The lottery must clearly establish the outcomes as either prize divisions or different lotteries. The phrases, *type of play* and *type of bet* are common terms used to categorize different lotteries.

For example, if you are asked to select any of several different play types (or bet types), like *straight, box, front pair, back pair* etc., then every selection is a different lottery. For every different type of bet you choose you are selecting a different type of lottery. But there could be also the instance where you are not asked to select a bet type, and the so called *straight* and *box* matches are division prizes of that lottery.

Draws with replacement have basically two variants: *straight* (exact order) and *box* (any order). They are primarily originated from the following lotteries:

Lotto 3/10

Three balls are drawn, one from each of three ball drawing machines, each containing ten balls numbered zero to nine.

This is the equivalent of drawing one ball three times with replacement from one draw machine containing ten balls numbered zero to nine.

Lotto 4/10

Four balls are drawn, one from each of four ball drawing machines, each containing ten balls numbered zero to nine.

This is the equivalent of drawing one ball four times with replacement from one draw machine containing ten balls numbered zero to nine.

Lotto 5/10

Five balls are drawn, one from each of five ball drawing machines, each containing ten balls numbered zero to nine.

This is the equivalent of drawing one ball five times with replacement from one draw machine containing ten balls numbered zero to nine.

Lotto 6/10

Six balls are drawn, one from each of six ball drawing machines, each containing ten balls numbered zero to nine.

This is the equivalent of drawing one ball six times with replacement from one draw machine containing ten balls numbered zero to nine.

3/10

Three balls are drawn, one from each of three lottery ball drawing machines, each containing ten balls numbered from 0 to 9.

Prize Structure and Probabilities		Favorable Outcomes	Probabilities
			1 in
Division 1:	Match the winning numbers in the exact order drawn. **(Straight)**	1	1,000
Division 2:	Match the winning numbers in any order drawn, with one number repeated. **(3-way box)**	3	333
Division 3:	Match the winning numbers in any order drawn. **(6-way box)**	6	167

Total Possible Outcomes

The number of total possible permutations when drawing three out of ten numbers with replacement is:

$$P(n,r) = n^r = P(10,3) = 10^3 = 1,000$$

Favorable Outcomes

<u>Straight (ABC)</u>

$$P(3,0) = 3!/(3 - 0)! = 3!/3! = 1$$

<u>3-way box (AAB)</u>

$$P(n,r) = n!/(x! \times y! \times z!) = P(3,2) = 3!/(2! \times 1!) = (3 \times 2)/2 = 3 \text{ (See \underline{note} below)}$$

6-way box (CAB)

$$P(3,3) = 3!/(3 - 3)! = 3!/0! = (3 \times 2)/1 = 6$$

<u>NOTE</u>: The outcomes when there are identical elements within the set are calculated with the formula:

$$P(n,r) = \frac{n!}{x! \times y! \times z!}$$

Where:
r = set of elements taken at a time
n = total number of elements
x = number of identical elements A
y = number of identical elements B
z = number of identical elements C

Lotteries of this type are played in
Arizona, , California, Connecticut, and Georgia.

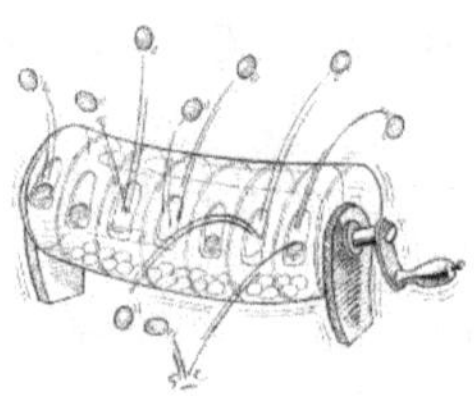

4/10

Prize Structure and Probabilities		Favorable Outcomes	Probabilities
			1 in
Division 1: Match all four winning numbers in the exact order drawn. **(Straight)**		1	10,000
Division 2: Match the winning numbers in any order drawn, with one number appearing three times. **(4-way box)**		4	2,500
Division 3: Match the winning numbers in any order drawn, with two numbers repeated. **(6-way box)**		6	1,667
Division 4: Match the winning numbers in any order drawn, with one number repeated. **(12-way box)**		12	833
Division 5: Match all the winning numbers in any order drawn. **(24-way box)**		24	417

Total Possible Outcomes

The number of total possible permutations when drawing four out of ten numbers with replacement is:

$$P(n,r) = n^r = P(10,4) = 10^4 = 10,000$$

Favorable Outcomes

Straight (ABCD)

$$P(4,0) = 4!/(4 - 0)! = 4!/4! = 1$$

4-way box (AAAB)

$$P(n,r) = n!/(x! \times y! \times z!) = P(4,2) = 4!/(3! \times 1!) = (4 \times 3 \times 2)/(3 \times 2) = 4$$

6-way box (AABB)

$$P(n,r) = n!/(x! \times y! \times z!) = P(4,2) = 4!/(2! \times 2!) = (4 \times 3 \times 2)/(2 \times 2) = 6$$

12-way box (AABC)

$$P(n,r) = n!/(x! \times y! \times z!) = P(4,3) = 4!/(2! \times 1! \times 1!) = (4 \times 3 \times 2)/(2 \times 1) = 12$$

24-way box (DCBA)

$$P(4,4) = 4!/(4 - 4)! = 4!/0! = 4 \times 3 \times 2 = 24$$

Lotteries of this type are played in
Arizona, , California, Connecticut, and Georgia (USA).

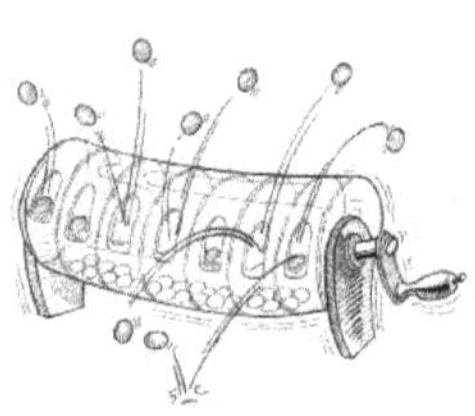

5/10

Prize Structure and Probabilities		Favorable Outcomes	Probabilities
			1 in
Division 1:	Match all five winning numbers in the exact order drawn. **(Straight)**	1	100,000
Division 2:	Match the winning numbers in any order drawn, with one number appearing four times. **(5-way box)**	5	20,000
Division 3:	Match the winning numbers in any order drawn, with one number appearing three times, and one number repeated. **(10-way box)**	10	10,000
Division 4:	Match the winning numbers in any order drawn, with one number appearing three times. **(20-way box)**	20	5,000
Division 5:	Match the winning numbers in any order drawn, with two numbers repeated. **(30-way box)**	30	3,333
Division 6:	Match the winning numbers in any order drawn, with one number repeated. **(60-way box)**	60	1,667
Division 7:	Match all the winning numbers in any order drawn. **(120-way box)**	120	833

Total Possible Outcomes

The number of total possible permutations when drawing five out of ten numbers with replacement is:

$$P(n,r) = n^r = P(10,5) = 10^5 = 100,000$$

Favorable Outcomes

Straight (ABCDE)

$$P(5,0) = 5!/(5 - 0)! = 5!/5! = 1$$

5-way box (AAAAB)

$$P(n,r) = n!/(x! \times y! \times z!) = P(5,2) = 5!/(4! \times 1!) = (5 \times 4 \times 3 \times 2)/(4 \times 3 \times 2) = 5$$

10-way box (AAABB)

$$P(n,r) = n!/(x! \times y! \times z!) = P(5,2) = 5!/(3! \times 2!) = (5 \times 4 \times 3 \times 2)/(3 \times 2 \times 2) = 10$$

20-way box (AAABC)

$$P(n,r) = n!/(x! \times y! \times z!) = P(5,3) = 5!/(3! \times 1! \times 1!) = (5 \times 4 \times 3 \times 2)/(3 \times 2) = 20$$

30-way box (AABBC)

$$P(n,r) = n!/(x! \times y! \times z!) = P(5,3) = 5!/(2! \times 2! \times 1!) = (5 \times 4 \times 3 \times 2)/(2 \times 2) = 30$$

60-way box (AABCD)

$$P(n,r) = n!/(x! \times y! \times z!) = P(5,4) = 5!/(2! \times 1! \times 1! \times 1!) = (5 \times 4 \times 3 \times 2)/2 = 60$$

120-way box (EDCBA)

$$P(5,5) = 5!/(5 - 5)! = 5!/0! = (5 \times 4 \times 3 \times 2)/1 = 120$$

A lottery of this type is played in Florida, USA.

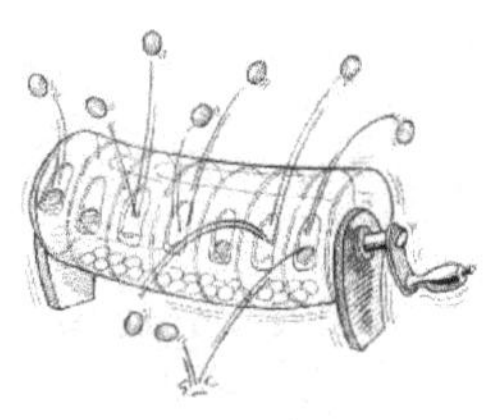

5/10 — Exact order
Prize Structure A

Prize Structure and Probabilities		Favorable Outcomes	Probabilities
			1 in
Division 1:	Match all five winning numbers in the exact order drawn.	1	100,000
Division 2:	Match the first four winning numbers, **or** the last four winning numbers in the exact order drawn.	18	5,556
Division 3:	Match the first three winning numbers, **or** the last three winning numbers in the exact order drawn.	180	556
Division 4:	Match the first two winning numbers, **or** the last two winning numbers in the exact order drawn.	1,791	56
Division 5:	Match the first winning number, **or** the last winning number in the exact order drawn.	17,010	6

Total Possible Outcomes

The number of total possible permutations when drawing five out of ten numbers with replacement is:

$$P(n,r) = n^r = P(10,5) = 10^5 = 100,000$$

Favorable Outcomes

Division 1

A B C D E	$P(n,r) = P(5,0) =$	1
	Favorable outcomes =	**1**

Division 2

A	B	C	D	Z	$P(n,r) = P(9,1) =$	9
V	B	C	D	E	$P(n,r) = P(9,1) =$	9
					Favorable outcomes =	**18**

Division 3

A	B	C	Y	Z	$P(n,r) = P(9,2) =$	81
A	B	C	Y	E	$P(n,r) = P(9,1) =$	9
A	W	C	D	E	$P(n,r) = P(9,1) =$	9
V	W	C	D	E	$P(n,r) = P(9,2) =$	81
					Favorable outcomes =	**180**

Division 4

A	B	X	D	E	$P(n,r) = P(9,1) =$	9
A	B	X	Y	E	$P(n,r) = P(9,2) =$	81
A	B	X	Y	Z	$P(n,r) = P(9,3) =$	729
A	B	X	D	Z	$P(n,r) = P(9,2) =$	81
A	W	X	D	E	$P(n,r) = P(9,2) =$	81
V	W	X	D	E	$P(n,r) = P(9,3) =$	729
V	B	X	D	E	$P(n,r) = P(9,2) =$	81
					Favorable outcomes =	**1,791**

Division 5

A	W	**C**	**D**	Z	$P(n,r) = P(9,2) =$	81
A	W	**C**	Y	**E**	$P(n,r) = P(9,2) =$	81
A	W	C	Y	Z	$P(n,r) = P(9,3) =$	729
A	W	X	**D**	Z	$P(n,r) = P(9,3) =$	729
A	W	X	Y	Z	$P(n,r) = P(9,4) =$	6,561
A	W	X	Y	**E**	$P(n,r) = P(9,3) =$	729
V	**B**	X	Y	**E**	$P(n,r) = P(9,3) =$	729
V	**B**	**C**	Y	**E**	$P(n,r) = P(9,2) =$	81
V	W	**C**	Y	**E**	$P(n,r) = P(9,3) =$	729
V	W	X	Y	**E**	$P(n,r) = P(9,4) =$	6,561

Favorable outcomes = 17,010

Remaining permutations

V	**B**	**C**	**D**	Z	$P(n,r) = P(9,2) =$	81
V	**B**	**C**	Y	Z	$P(n,r) = P(9,3) =$	729
V	**B**	X	**D**	Z	$P(n,r) = P(9,3) =$	729
V	**B**	X	Y	Z	$P(n,r) = P(9,4) =$	6,561
V	W	**C**	**D**	Z	$P(n,r) = P(9,3) =$	729
V	W	**C**	Y	Z	$P(n,r) = P(9,4) =$	6,561
V	W	X	**D**	Z	$P(n,r) = P(9,4) =$	6,561
V	W	X	Y	Z	$P(n,r) = P(9,5) =$	59,049

Favorable outcomes = 81,000

Summary

All five	1 × P$(9,0)$ = 1 × 1	=	1	
				1
First (or last) four	2 × P$(9,1)$ = 2 × 9	=	18	
				18
First (or last) three	2 × P$(9,1)$ = 2 × 9	=	18	
	2 × P$(9,2)$ = 2 × 81	=	162	
				180
First (or last) two	1 × P$(9,1)$ = 1 × 9	=	9	
	4 × P$(9,2)$ = 4 × 81	=	324	
	2 × P$(9,3)$ = 2 × 729	=	1,458	
				1,791
First (or last) one	3 × P$(9,2)$ = 3 × 81	=	243	
	5 × P$(9,3)$ = 5 × 729	=	3,645	
	2 × P$(9,4)$ = 2 × 6,561	=	13,122	
				17,010
Other permutations	1 × P$(9,2)$ = 1 × 81	=	81	
	3 × P$(9,3)$ = 3 × 729	=	2,187	
	3 × P$(9,4)$ = 3 × 6,561	=	19,683	
	1 × P$(9,5)$ = 1 × 59,049	=	59,049	
				81,000
	Total possible outcomes:			100,000

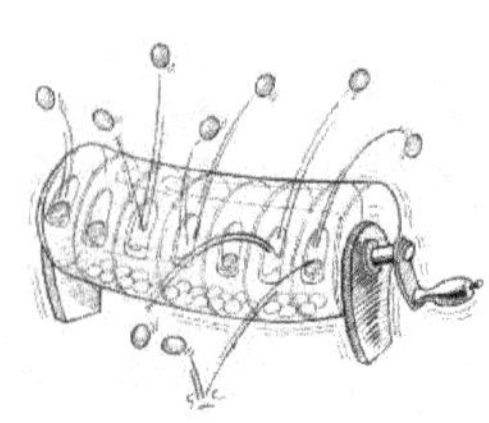

5/10 — Exact order
Prize Structure B

Prize Structure and Probabilities		Favorable Outcomes	Probabilities
			1 in
Division 1:	Match all five winning numbers in the exact order drawn.	1	100,000
Division 2:	Match the first four winning numbers, **or** the last four winning numbers in the exact order drawn.	18	5,556
Division 3:	Match the first three winning numbers, **and** the last number in the exact order drawn.	9	11,111
Division 4:	Match the first winning number, **and** the last three winning numbers in the exact order drawn.	9	11,111
Division 5:	Match the first three winning numbers, **or** the last three winning numbers in the exact order drawn.	162	617
Division 6:	Match the first two winning numbers, **and** the last two winning numbers in the exact order drawn.	9	11,111
Division 7:	Match the first two winning numbers, **and** the last winning number in the exact order drawn.	81	1,235
Division 8:	Match the first winning number, **and** the last two winning numbers in the exact order drawn.	81	1,235
Division 9:	Match the first two winning numbers, **or** the last two winning numbers in the exact order drawn.	1,620	62
Division 10:	Match the first winning number, **and** the last winning number in the exact order drawn.	810	123
Division 11:	Match the first winning number, **or** the last winning number in the exact order drawn.	16,200	6

Total Possible Outcomes

The number of total possible permutations when drawing five out of ten numbers with replacement is:

$$P(n,r) = n^r = P(10,5) = 10^5 = 100,000$$

Favorable Outcomes

Division 1

A	B	C	D	E		
					$P(n,r) = P(5,0) =$	1
					Favorable outcomes =	**1**

Division 2

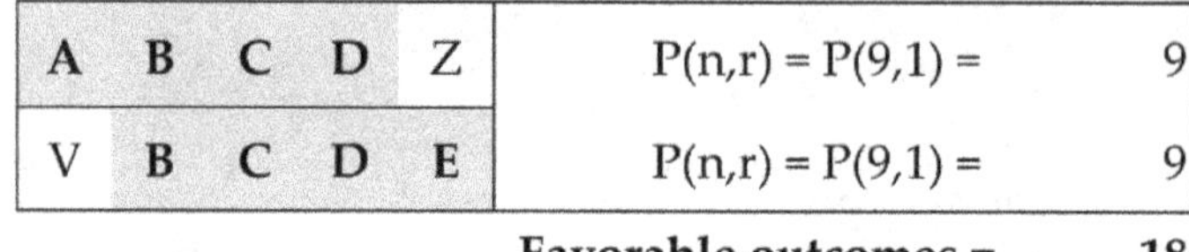

A	B	C	D	Z	$P(n,r) = P(9,1) =$	9
V	B	C	D	E	$P(n,r) = P(9,1) =$	9
					Favorable outcomes =	**18**

Division 3

A	B	C	Y	E	$P(n,r) = P(9,1) =$	9
					Favorable outcomes =	9

Division 4

A	W	C	D	E	$P(n,r) = P(9,1) =$	9
					Favorable outcomes =	9

Division 5

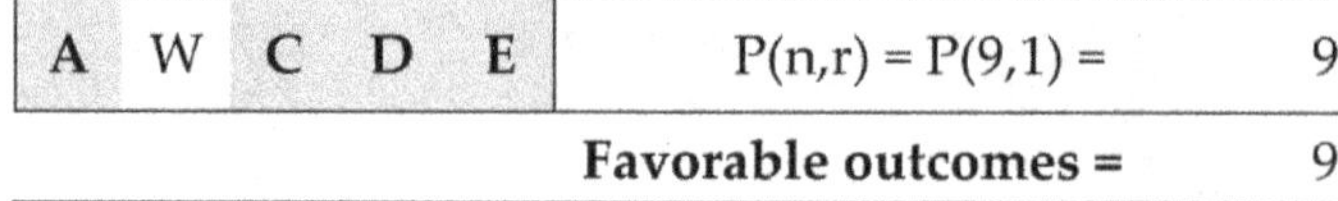

A	B	C	Y	Z	$P(n,r) = P(9,2) =$	81
V	W	C	D	E	$P(n,r) = P(9,2) =$	81
					Favorable outcomes =	**162**

Division 6

A	B	X	D	E	$P(n,r) = P(9,1) =$	9
					Favorable outcomes =	9

Division 7

A	B	X	Y	E	P(n,r) = P(9,2) =	81
					Favorable outcomes =	**81**

Division 8

A	W	X	D	E	P(n,r) = P(9,2) =	81
					Favorable outcomes =	**81**

Division 9

A	B	X	Y	Z	P(n,r) = P(9,3) =	729
A	B	X	D	Z	P(n,r) = P(9,2) =	81
V	W	X	D	E	P(n,r) = P(9,3) =	729
V	B	X	D	E	P(n,r) = P(9,2) =	81
					Favorable outcomes =	**1,620**

Division 10

A	W	C	Y	E	P(n,r) = P(9,2) =	81
A	W	X	Y	E	P(n,r) = P(9,3) =	729
					Favorable outcomes =	**810**

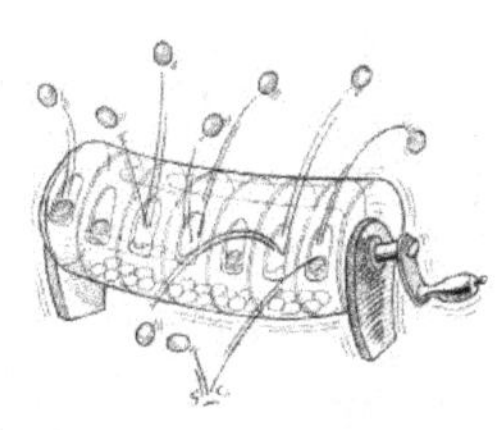

Division 11

A	W	C	D	Z	P(n,r) = P(9,2) =	81
A	W	C	Y	Z	P(n,r) = P(9,3) =	729
A	W	X	D	Z	P(n,r) = P(9,3) =	729
A	W	X	Y	Z	P(n,r) = P(9,4) =	6,561
V	B	X	Y	E	P(n,r) = P(9,3) =	729
V	B	C	Y	E	P(n,r) = P(9,2) =	81
V	W	C	Y	E	P(n,r) = P(9,3) =	729
V	W	X	Y	E	P(n,r) = P(9,4) =	6,561

Favorable outcomes = 16,200

Remaining permutations

V	B	C	D	Z	P(n,r) = P(9,2) =	81
V	B	C	Y	Z	P(n,r) = P(9,3) =	729
V	B	X	D	Z	P(n,r) = P(9,3) =	729
V	B	X	Y	Z	P(n,r) = P(9,4) =	6,561
V	W	C	D	Z	P(n,r) = P(9,3) =	729
V	W	C	Y	Z	P(n,r) = P(9,4) =	6,561
V	W	X	D	Z	P(n,r) = P(9,4) =	6,561
V	W	X	Y	Z	P(n,r) = P(9,5) =	59,049

Favorable outcomes = 81,000

A lottery of this type is played in Georgia, USA.

6/10 — Exact order

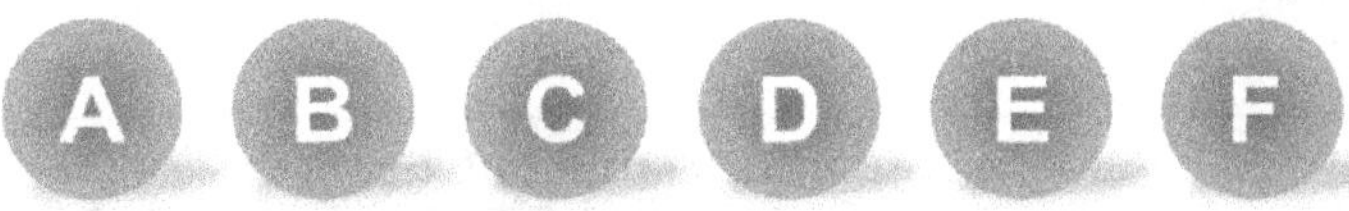

Prize Structure and Probabilities

		Favorable Outcomes	Probabilities
			1 in
Division 1:	Match all six winning numbers in the exact order drawn.	1	1,000,000
Division 2:	Match the first five winning numbers, **or** the last five winning numbers in the exact order drawn.	18	55,556
Division 3:	Match the first four winning numbers, **or** the last four winning numbers in the exact order drawn.	180	5,556
Division 4:	Match the first three winning numbers, **or** the last three winning numbers in the exact order drawn.	1,800	556
Division 5:	Match the first two winning numbers, **or** the last two winning numbers in the exact order drawn.	17,901	56
Division 6:	Match the first winning number, **or** the last winning number in the exact order drawn.	170,100	6

Total Possible Outcomes

The number of total possible permutations when drawing five out of ten numbers with replacement is:

$$P(n,r) = n^r = P(10,6) = 10^6 = 1,000,000$$

Favorable Outcomes

Division 1

A B C D E F	$P(n,r) = P(6,0) =$	1
	Favorable outcomes =	**1**

Division 2

A	B	C	D	E	Z	$P(n,r) = P(9,1) =$	9
U	B	C	D	E	F	$P(n,r) = P(9,1) =$	9

Favorable outcomes = 18

Division 3

A	B	C	D	Y	F	$P(n,r) = P(9,1) =$	9
A	B	C	D	Y	Z	$P(n,r) = P(9,2) =$	81
A	V	C	D	E	F	$P(n,r) = P(9,1) =$	9
U	V	C	D	E	F	$P(n,r) = P(9,2) =$	81

Favorable outcomes = 180

Division 4

A	B	C	X	E	F	$P(n,r) = P(9,1) =$	9
A	B	C	X	Y	F	$P(n,r) = P(9,2) =$	81
A	B	C	X	E	Z	$P(n,r) = P(9,2) =$	81
A	B	C	X	Y	Z	$P(n,r) = P(9,3) =$	729
A	B	W	D	E	F	$P(n,r) = P(9,1) =$	9
A	V	W	D	E	F	$P(n,r) = P(9,2) =$	81
U	B	W	D	E	F	$P(n,r) = P(9,2) =$	81
U	V	W	D	E	F	$P(n,r) = P(9,3) =$	729

Favorable outcomes = 1,800

Division 5

						P(n,r)	
A	B	W	D	E	Z	P(n,r) = P(9,2) =	81
A	B	W	D	Y	F	P(n,r) = P(9,2) =	81
A	B	W	D	Y	Z	P(n,r) = P(9,3) =	729
A	B	W	X	E	Z	P(n,r) = P(9,3) =	729
A	B	W	X	Y	F	P(n,r) = P(9,3) =	729
A	B	W	X	Y	Z	P(n,r) = P(9,4) =	6,561
A	B	W	X	Y	F	P(n,r) = P(9,2) =	81
A	V	C	X	E	F	P(n,r) = P(9,2) =	81
A	V	W	X	E	F	P(n,r) = P(9,3) =	729
U	B	C	X	E	F	P(n,r) = P(9,2) =	81
U	B	W	X	E	F	P(n,r) = P(9,3) =	729
U	V	C	X	E	F	P(n,r) = P(9,3) =	729
U	V	W	X	E	F	P(n,r) = P(9,4) =	6,561

Favorable outcomes =   **17,901**

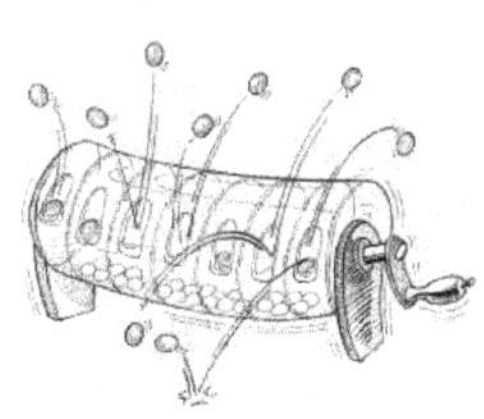

Division 6

A	V	C	D	E	Z	P(n,r) = P(9,2) =		81
A	V	C	D	Y	Z	P(n,r) = P(9,3) =		729
A	V	C	X	E	Z	P(n,r) = P(9,3) =		729
A	V	C	X	Y	Z	P(n,r) = P(9,4) =		6,561
A	V	W	D	E	Z	P(n,r) = P(9,3) =		729
A	V	W	D	Y	Z	P(n,r) = P(9,4) =		6,561
A	V	W	X	Y	Z	P(n,r) = P(9,5) =		59,049
A	V	W	X	E	Z	P(n,r) = P(9,4) =		6,561
U	B	C	X	Y	F	P(n,r) = P(9,3) =		729
U	B	C	D	Y	F	P(n,r) = P(9,2) =		81
U	B	W	D	Y	F	P(n,r) = P(9,3) =		729
U	V	C	D	Y	F	P(n,r) = P(9,3) =		729
U	V	C	X	Y	F	P(n,r) = P(9,4) =		6,561
U	V	W	D	Y	F	P(n,r) = P(9,4) =		6,561
U	V	W	X	Y	F	P(n,r) = P(9,5) =		59,049
U	B	W	X	Y	F	P(n,r) = P(9,4) =		6,561
A	V	C	D	Y	F	P(n,r) = P(9,2) =		81
A	V	C	X	Y	F	P(n,r) = P(9,3) =		729
A	V	W	D	Y	F	P(n,r) = P(9,3) =		729
A	V	W	X	Y	F	P(n,r) = P(9,4) =		6,561

Favorable outcomes = 170,100

Remaining permutations

U	**B**	**C**	**D**	**E**	Z	$P(n,r) = P(9,2) =$	81
U	**B**	**C**	**D**	Y	Z	$P(n,r) = P(9,3) =$	729
U	**B**	**C**	X	**E**	Z	$P(n,r) = P(9,3) =$	729
U	**B**	**C**	X	Y	Z	$P(n,r) = P(9,4) =$	6,561
U	**B**	W	**D**	**E**	Z	$P(n,r) = P(9,3) =$	729
U	**B**	W	**D**	Y	Z	$P(n,r) = P(9,4) =$	6,561
U	**B**	W	X	**E**	Z	$P(n,r) = P(9,4) =$	6,561
U	**B**	W	X	Y	Z	$P(n,r) = P(9,5) =$	59,049
U	V	**C**	**D**	Y	Z	$P(n,r) = P(9,4) =$	6,561
U	V	**C**	**D**	E	Z	$P(n,r) = P(9,3) =$	729
U	V	**C**	X	E	Z	$P(n,r) = P(9,4) =$	6,561
U	V	**C**	X	Y	Z	$P(n,r) = P(9,5) =$	59,049
U	V	W	**D**	Y	Z	$P(n,r) = P(9,5) =$	59,049
U	V	W	**D**	E	Z	$P(n,r) = P(9,4) =$	6,561
U	V	W	X	**E**	Z	$P(n,r) = P(9,5) =$	59,049
U	V	W	X	Y	Z	$P(n,r) = P(9,6) =$	531,441

Favorable outcomes = 810,000

A lottery of this type is played in Australia,
with only five divisions.

Summary

All six	$1 \times P_{(9,0)} = 1 \times 1$	$=$	1	
				1
First (or last) five	$2 \times P_{(9,1)} = 2 \times 9$	$=$	18	
				18
First (or last) four	$2 \times P_{(9,1)} = 2 \times 9$	$=$	18	
	$2 \times P_{(9,2)} = 2 \times 81$	$=$	162	
				180
First (or last) three	$2 \times P_{(9,1)} = 2 \times 9$	$=$	18	
	$4 \times P_{(9,2)} = 4 \times 81$	$=$	324	
	$2 \times P_{(9,3)} = 2 \times 729$	$=$	1,458	
				1,800
First (or last) two	$5 \times P_{(9,2)} = 5 \times 81$	$=$	405	
	$6 \times P_{(9,3)} = 6 \times 729$	$=$	4,374	
	$2 \times P_{(9,4)} = 2 \times 6,561$	$=$	13,122	
				17,901
First (or last) one	$3 \times P_{(9,2)} = 3 \times 81$	$=$	243	
	$8 \times P_{(9,3)} = 8 \times 729$	$=$	5,832	
	$7 \times P_{(9,4)} = 7 \times 6,561$	$=$	45,927	
	$2 \times P_{(9,5)} = 2 \times 59,049$	$=$	118,098	
				170,100
Other permutations	$1 \times P_{(9,2)} = 1 \times 81$	$=$	81	
	$4 \times P_{(9,3)} = 4 \times 729$	$=$	2,916	
	$6 \times P_{(9,4)} = 6 \times 6,561$	$=$	39,366	
	$4 \times P_{(9,5)} = 4 \times 59,049$	$=$	236,196	
	$1 \times P_{(9,6)} = 1 \times 531,441$	$=$	531,441	
				810,000
	Total possible outcomes:			1,000,000

CHAPTER 5

DRAWS WITHOUT REPLACEMENT
ONE DRAW MACHINE

Draws without replacement include lotteries in which every number drawn is not returned into the system before drawing the next number. As it could be expected, this procedure does not produce repeated numbers. Because order is not important in this case, most of these problems are solved by using the formulas applied to combinations.

In these draws, the player selects five, six, or seven numbers from one to a certain maximum specified by the lottery. They are usually denoted by the abbreviation n/N, which means that you must select n numbers from 1 to N.

A typical example is 6/45, where you select six numbers from 1 to 45. In some cases, in addition to the n numbers, one or two more numbers are drawn from the same set. Those are supplementary or supplementary numbers.

NOTE:

Some lotteries offer an additional number known as a supplementary or supplementary number. In this book, we will use the abbreviation "suppl." to refer to this extra number.

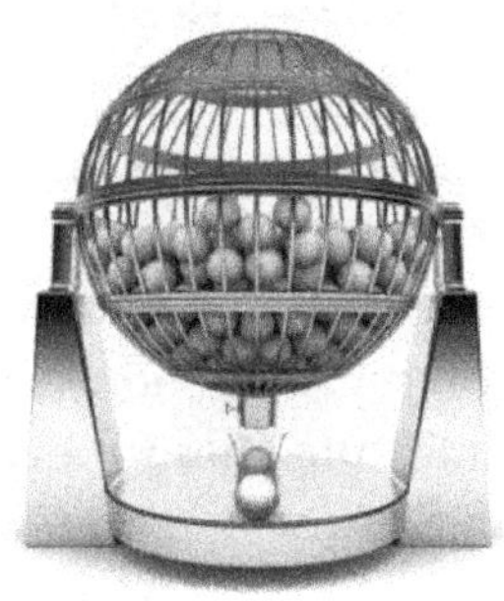

4/40 — Exact Order

A lottery ball drawing machine contains forty balls numbered from 1 to 40. Four balls are drawn without replacement. These are the winning numbers.

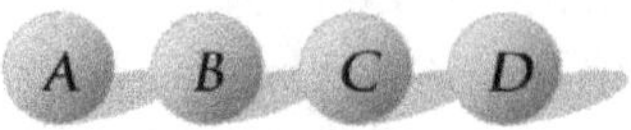

WINNING NUMBERS

Prize Structure and Probabilities		Favorable Outcomes	Probabilities
			1 in
Division 1:	Match all four winning numbers in the exact order drawn (4-straight).	1	2,193,360
Division 2:	Match any three winning numbers in the exact order drawn (3-straight).	144	15,232
Division 3:	Match any two winning numbers in the exact order drawn (2-straight).	7,998	274
Division 4:	Match any winning number in the exact order drawn (1-straight).	202,904	11

Total Possible Outcomes

The number of permutations without repetitions when drawing 4 out of 40 numbers is:

$$P(40,4) = 40 \times 39 \times 38 \times 37 = 2,193,360$$

Favorable Outcomes and Probabilities

The letters WXYZ represent any 4 of the 36 nonwinning numbers (any of the letters WXYZ represents any nonwinning number).

To count the favorable outcomes for every division we divide the total possible outcomes in five cases:

Case 1	The set of four numbers includes all the winning numbers: **ABCD**
Case 2	The set of four numbers includes three winning numbers and one nonwinning number: **ABC**Z
Case 3	The set of four numbers includes two winning numbers and two nonwinning numbers: **AB**YZ
Case 4	The set of four numbers includes one winning number and three nonwinning numbers: **A**XYZ
Case 5	The set of four numbers includes any four nonwinning numbers: WXYZ

Let us consider each case separately:

Case 1:	The set of four numbers includes all the winning numbers: **ABCD**
	The number of permutations without repetition of four numbers in four places is P(4,4) = 4 x 3 x 2 x 1 = 24
	The following are all the 24 possible arrangements of **ABCD**:

A	B	C	D	4-straight

A	**B**	D	C	2-straight
A	D	**C**	B	2-straight
A	C	B	**D**	2-straight
D	**B**	**C**	A	2-straight
C	**B**	A	**D**	2-straight
B	A	**C**	**D**	2-straight

A	C	D	B	1-straight
A	D	B	C	1-straight
C	**B**	D	A	1-straight
D	**B**	A	C	1-straight
B	D	**C**	A	1-straight
D	A	**C**	B	1-straight
B	C	A	**D**	1-straight
C	A	B	**D**	1-straight

Other permutations:

B	A	D	C
B	C	D	A
B	D	A	C
C	A	D	B
C	D	A	B
C	D	B	A
D	A	B	C
D	C	A	B
D	C	B	A

Different possible arrangements within **ABCD**:

Division 1 (4-straight)	1
Division 2 (3-straight)	0
Division 3 (2-straight)	6
Division 4 (1-straight)	8
Other arrangements	9
Sum:	24

Case 2: The set of four numbers includes three winning numbers and one nonwinning number: **ABCZ**

The number of permutations without repetition of three numbers in four places is: $P(4,3) = 4 \times 3 \times 2 = 24$

The following are all the 24 possible arrangements of **ABCZ**:

A	**B**	**C**	Z	3-straight
A	**B**	Z	C	2-straight
A	Z	**C**	B	2-straight
Z	**B**	**C**	A	2-straight
A	C	B	Z	1-straight
A	C	Z	B	1-straight
A	Z	B	C	1-straight
C	**B**	A	Z	1-straight
C	**B**	Z	A	1-straight
Z	**B**	A	C	1-straight

B	A	**C**	Z	1-straight
B	Z	**C**	A	1-straight
Z	A	**C**	B	1-straight

Other permutations

B	A	Z	C
B	C	Z	A
B	Z	A	C
C	A	Z	B
C	Z	A	B
C	Z	B	A
Z	A	B	C
Z	C	A	B
Z	C	B	A
B	C	A	Z
C	A	B	Z

However, this is a partial result, because there are 36 nonwinning numbers, represented by the letter Z. We also must calculate the different possible arrangements within **ABC**Z.

<u>Different possible arrangements within **ABC**Z:</u>

The different possible <u>combinations</u> that can be made with one number (Z) in four positions (**ABC**Z) are C(4,1) = 4

ABCZ	ABZC	AZBC	ZABC

The different possible <u>permutations</u> that can be made with any nonwinning number Z are P(36,1) = 36

Then, the total of different arrangements that can be made with **ABC**Z is C(4,1) x P(36,1) = 4 x 36 = 144

And therefore, we must multiply each one of all possible arrangements of **ABC**Z by 144:

Division 1 (4-straight)		0
Division 2 (3-straight)	1 x 144 =	144
Division 3 (2-straight)	3 x 144 =	432
Division 4 (1-straight)	9 x 144 =	1,296
Other arrangements	11 x 144 =	1,584
	Sum:	3,456

Case 3: The set of four numbers includes two winning numbers and two nonwinning numbers: **AB**YZ.

The number of permutations without repetitions of two numbers in four places is: $P(4,2) = 4 \times 3 = 12$

The following are all the 12 possible arrangements of **AB**YZ:

A	**B**	Y	Z	2-straight

A	Y	B	Z	1-straight
A	Y	Z	B	1-straight
Y	**B**	A	Z	1-straight
Y	**B**	Z	A	1-straight

Other permutations

B	A	Y	Z
B	Y	A	Z
B	Y	Z	A
Y	A	B	Z
Y	A	Z	B
Y	Z	A	B
Y	Z	B	A

However, this is a partial result, because there are 36 nonwinning numbers, represented by either of the letters YZ. We also must calculate the different possible arrangements within **AB**YZ.

<u>Different possible arrangements within **AB**YZ:</u>

The different possible <u>combinations</u> that can be made with two numbers (YZ) in four positions (**AB**YZ) are $C(4,2) = 6$

The different possible permutations that can be made with any two nonwinning numbers YZ are P(36,2) = 1,260

Then, the total of different arrangements that can be made with **AB**YZ is C(4,2) x P(36,2) = 6 x 1,260 = 7,560

And therefore, we must multiply each one of all possible arrangements of **AB**YZ by 7,560:

Division 1 (4-straight)		0
Division 2 (3-straight)		0
Division 3 (2-straight)	1 x 7,560 =	7,560
Division 4 (1-straight)	4 x 7,560 =	30,240
Other arrangements	7 x 7,560 =	52,920
	Sum:	90,720

Case 4: The set of four numbers includes one winning number and three nonwinning numbers: **A**XYZ

The number of permutations without repetitions of one number in four places is P(4,1) = 4

The following are all 4 possible arrangements of **A**XYZ:

A	X	Y	Z	1-straight

Other permutations

X	A	Y	Z
X	Y	A	Z
X	Y	Z	A

However, this is a partial result, because there are 36 nonwinning numbers, represented by either of the letters XYZ. We also must calculate the different possible arrangements within **A**XYZ.

<u>Different possible arrangements within **A**XYZ:</u>

The different possible <u>combinations</u> that can be made with three numbers (XYZ) in four positions (**A**XYZ) are C(4,3) = 4

The different possible permutations that can be made with any three nonwinning numbers XYZ are P(36,3) = 42,840

Then, the total of different arrangements that can be made with **A**XYZ is C(4,3) x P(36,3) = 4 x 42,840 = 171,360

And therefore, we must multiply each one of all possible arrangements of **A**XYZ by 171,360:

Division 1 (4-straight)		0
Division 2 (3-straight)		0
Division 3 (2-straight)		0
Division 4 (1-straight)	1 x 171,360 =	171,360
Other arrangements	3 x 171,360 =	514,080
	Sum:	685,440

Case 5: The set of four numbers includes four nonwinning numbers: WXYZ.

All possible arrangements of WXYZ: P(36,4) = 1,413,720

Summary

	ABCD	ABCZ	ABYZ	AXYZ	WXYZ	Favorable Outcomes
Division 1 (4-straight)	1					1
Division 2 (3-straight)	0	144				144
Division 3 (2-straight)	6	432	7,560			7,998
Division 4 (1-straight)	8	1,296	30,240	171,360		202,904
Non-winning arrangements	9	1,584	52,920	514,080	1,413,720	1,982,313
Sum	24	3,456	90,720	685,440	1,413,720	
Total Possible Outcomes:						2,193,360

A lottery of this type is played in New Zealand.

4/45 — Exact Order

A lottery ball drawing machine contains forty-five balls numbered from 1 to 45. Four balls are drawn without replacement. These are the winning numbers.

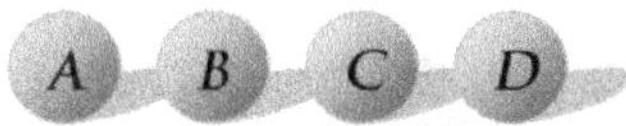

WINNING NUMBERS

Prize Structure and Probabilities		Favorable Outcomes	Probabilities
			1 in
Division 1:	Match all four winning numbers in the exact order drawn (4-straight).	1	3,575,880
Division 2:	Match any three winning numbers in the exact order drawn (3-straight).	164	21,804
Division 3:	Match any two winning numbers in the exact order drawn (2-straight).	10,338	346
Division 4:	Match any winning number in the exact order drawn (1-straight).	296,684	12

Total Possible Outcomes

The number of possible permutations without repetitions when drawing 4 out of 45 numbers is:

$$P(45,4) = 45 \times 44 \times 43 \times 42 = 3,575,880$$

Favorable Outcomes and Probabilities

The letters WXYZ represent any 4 of the 41 nonwinning numbers (any of the letters WXYZ represents any nonwinning number).

To count the favorable outcomes for every division we divide the total possible outcomes in five cases:

Case 1	The set of four numbers includes all the winning numbers: **ABCD**
Case 2	The set of four numbers includes three winning numbers and one nonwinning number: **ABC**Z
Case 3	The set of four numbers includes two winning numbers and two nonwinning numbers: **AB**YZ
Case 4	The set of four numbers includes one winning number and three nonwinning numbers: **A**XYZ
Case 5	The set of four numbers includes any four nonwinning numbers: WXYZ

Let us consider each case separately:

Case 1:	The set of four numbers includes all the winning numbers: **ABCD**
	The number of permutations without repetition of four numbers in four places is P(4,4) = 4 x 3 x 2 x 1 = 24
	The following are all the 24 possible arrangements of **ABCD**:

A	**B**	**C**	**D**	4-straight
A	**B**	D	C	2-straight
A	D	**C**	B	2-straight
A	C	B	**D**	2-straight
D	**B**	**C**	A	2-straight
C	**B**	A	**D**	2-straight
B	A	**C**	**D**	2-straight
A	C	D	B	1-straight
A	D	B	C	1-straight
C	**B**	D	A	1-straight
D	**B**	A	C	1-straight
B	D	**C**	A	1-straight
D	A	**C**	B	1-straight
B	C	A	**D**	1-straight
C	A	B	**D**	1-straight

Other permutations

B	A	D	C
B	C	D	A
B	D	A	C
C	A	D	B
C	D	A	B
C	D	B	A
D	A	B	C
D	C	A	B
D	C	B	A

<u>Different possible arrangements within **ABCD**:</u>

Division 1 (4-straight)	1
Division 2 (3-straight)	0
Division 3 (2-straight)	6
Division 4 (1-straight)	8
Other arrangements	9
Sum:	24

Case 2: The set of four numbers includes three winning numbers and one nonwinning number: **ABCZ**

The number of permutations without repetition of three numbers in four places is: P(4,3) = 4 x 3 x 2 = 24

The following are all the 24 possible arrangements of **ABCZ**:

A	**B**	**C**	Z	3-straight
A	**B**	Z	C	2-straight
A	Z	**C**	B	2-straight
Z	**B**	**C**	A	2-straight
A	C	B	Z	1-straight
A	C	Z	B	1-straight
A	Z	B	C	1-straight
C	**B**	A	Z	1-straight
C	**B**	Z	A	1-straight
Z	**B**	A	C	1-straight

B	A	**C**	Z	1-straight
B	Z	**C**	A	1-straight
Z	A	**C**	B	1-straight

Other permutations

B	A	Z	C
B	C	Z	A
B	Z	A	C
C	A	Z	B
C	Z	A	B
C	Z	B	A
Z	A	B	C
Z	C	A	B
Z	C	B	A
B	C	A	Z
C	A	B	Z

However, this is a partial result, because there are 41 nonwinning numbers, represented by the letter Z. We also must calculate the different possible arrangements within **ABCZ**.

<u>Different possible arrangements within **ABCZ**:</u>

The different possible <u>combinations</u> that can be made with one number (Z) in four positions (**ABCZ**) are C(4,1) = 4

ABCZ	ABZC	AZBC	ZABC

The different possible <u>permutations</u> that can be made with any nonwinning number Z are P(41,1) = 41

Then, the total of different possible arrangements that can be made with **ABCZ** is C(4,1) x P(41,1) = 4 x 41 = 164

And therefore, we must multiply each one of all possible arrangements of **ABCZ** by 164:

Division 1 (4-straight)		0
Division 2 (3-straight)	1 x 164 =	164
Division 3 (2-straight)	3 x 164 =	492
Division 4 (1-straight)	9 x 164 =	1,476
Other arrangements	11 x 164 =	1,804
	Sum:	3,936

Case 3: The set of four numbers includes two winning numbers and two nonwinning numbers: **AB**YZ.

The number of permutations without repetitions of two numbers in four places is: $P(4,2) = 4 \times 3 = 12$

The following are all the 12 possible arrangements of **AB**YZ:

A	B	Y	Z	2-straight
A	Y	B	Z	1-straight
A	Y	Z	B	1-straight
Y	B	A	Z	1-straight
Y	B	Z	A	1-straight

Other permutations

B	A	Y	Z
B	Y	A	Z
B	Y	Z	A
Y	A	B	Z
Y	A	Z	B
Y	Z	A	B
Y	Z	B	A

However, this is a partial result, because there 41 nonwinning numbers, represented by either of the letters YZ. We also must calculate the different possible arrangements within **AB**YZ.

Different possible arrangements within **AB**YZ:

The different possible <u>combinations</u> that can be made with two numbers (YZ) in four positions (**AB**YZ) are $C(4,2) = 6$

The different possible permutations that can be made with any two nonwinning numbers YZ are P(41,2) = 1,640

Then, the total of different possible arrangements that can be made with **AB**YZ is C(4,2) x P(41,2) = 6 x 1,640 = 9,840

And therefore, we must multiply each one of all possible arrangements of **AB**YZ by 9,840:

Division 1 (4-straight)		0
Division 2 (3-straight)		0
Division 3 (2-straight)	1 x 9,840 =	9,840
Division 4 (1-straight)	4 x 9,840 =	39,360
Other arrangements	7 x 9,840 =	68,880
	Sum:	118,080

Case 4: The set of four numbers includes one winning number and three nonwinning numbers: **A**XYZ

The number of permutations without repetitions of one number in four places is P(4,1) = 4

The following are all 4 possible arrangements of **A**XYZ:

A	X	Y	Z	1-straight

Other permutations

X	A	Y	Z
X	Y	A	Z
X	Y	Z	A

However, this is a partial result, because there are 41 nonwinning numbers, represented by either of the letters XYZ. We also must calculate the different possible arrangements within **A**XYZ.

<u>Different possible arrangements within **A**XYZ:</u>

The different possible <u>combinations</u> that can be made with three numbers (XYZ) in four positions (**A**XYZ) are C(4,3) = 4

The different possible permutations that can be made with any three nonwinning numbers XYZ are P(41,3) = 63,960

Then, the total of different possible arrangements that can be made with $\mathbf{A}XYZ$ is C(4,3) x P(41,3) = 4 x 63,960 = 255,840

And therefore, we must multiply each one of all possible arrangements of $\mathbf{A}XYZ$ by 255,840:

Division 1 (4-straight)		0
Division 2 (3-straight)		0
Division 3 (2-straight)		0
Division 4 (1-straight)	1 x 255,840 =	255,840
Other arrangements	3 x 255,840 =	767,520
	Sum:	1,023,360

Case 5: The set of four numbers includes four nonwinning numbers: WXYZ.

All possible arrangements of WXYZ: P(41,4) = 2,430,480

Summary

	ABCD	ABCZ	ABYZ	AXYZ	WXYZ	Favorable Outcomes
Division 1 (4-straight)	1					1
Division 2 (3-straight)	0	164				164
Division 3 (2-straight)	6	492	9,840			10,338
Division 4 (1-straight)	8	1,476	39,360	255,840		296,684
Non-winning arrangements	9	1,804	68,880	767,520	2,430,480	3,268,693
Sum	24	3,936	118,080	1,023,360	2,430,480	
Total Possible Outcomes:						3,575,880

A lottery of this type is played in New South Wales, Australia.

5/28

The lottery drawing machine contains twenty-eight balls numbered from 1 to 28. Five balls are drawn without replacement.
These are the winning numbers.

Prize Structure and Probabilities

5/28

	Match Winning Numbers	Favorable Outcomes	Probabilities
			1 in
DIV 1	5	1	98,280
DIV 2	4	115	855
DIV 3	3	2,530	39
DIV 4	2	17,710	6

Total Possible Outcomes: C(28,5) = 98,280

Calculation of Favorable Outcomes					
W = Winning numbers (5)	S = Suppl. numbers (0)	R = Remaining numbers (23)			
MATCH	W	S			
DIV 1	5	0	$C_5^5 \times C_0^0 \times C_0^{23}$ = 1 x 1 x 1 =		1
DIV 2	4	0	$C_4^5 \times C_0^0 \times C_1^{23}$ = 5 x 1 x 23 =		115
DIV 3	3	0	$C_3^5 \times C_0^0 \times C_2^{23}$ = 10 x 1 x 253 =		2,530
DIV 4	2	0	$C_2^5 \times C_0^0 \times C_3^{23}$ = 10 x 1 x 1,771 =		17,710

A lottery of this type is played in Mexico.

5/31 (1 supplementary number)

The lottery drawing machine contains thirty-one balls numbered from 1 to 31. Five balls are drawn without replacement.

These are the winning numbers. A sixth ball is drawn without replacement. This is the supplementary number.

Prize Structure and Probabilities

5/31

	Match			
	Winning Numbers	Suppl. Number	Favorable Outcomes	Probabilities
				1 in
DIV 1	5		1	169,911
DIV 2	4	1	5	33,982
DIV 3	4		125	1,359
DIV 4	3		3,250	52

Total Possible Outcomes: $C(31,5) = 169,911$

Calculation of Favorable Outcomes

W = Winning numbers (5)   S = Suppl. number (1)   R = Remaining numbers (25)

MATCH

	W	S	W		S		R		
DIV 1	5	0	C_5^5	$\times$	C_0^1	$\times$	C_0^{25}	$= 1 \times 1 \times 1 =$	1
DIV 2	4	1	C_4^5	$\times$	C_1^1	$\times$	C_0^{25}	$= 5 \times 1 \times 1 =$	5
DIV 3	4	0	C_4^5	$\times$	C_0^1	$\times$	C_1^{25}	$= 5 \times 1 \times 25 =$	125
DIV 4	3	1	C_3^5	$\times$	C_1^1	$\times$	C_1^{25}	$= 10 \times 1 \times 25 = 250$	3,250
	3	0	C_3^5	$\times$	C_0^1	$\times$	C_2^{25}	$= 10 \times 1 \times 300 = 3,000$	

A lottery of this type is played in Japan.

5/32

The lottery drawing machine contains thirty-two balls numbered from 1 to 32. Five balls are drawn without replacement.

These are the winning numbers.

Prize Structure and Probabilities

5/32

	Match Winning Numbers	Favorable Outcomes	Probabilities 1 in
DIV 1	5	1	201,376
DIV 2	4	135	1,492
DIV 3	3	3,510	57
DIV 4	2	29,250	7

Total Possible Outcomes: $C(32,5) = 201,376$

Calculation of Favorable Outcomes

W = Winning numbers (5)   S = Suppl. number (0)   R = Remaining numbers (27)

MATCH

	W	S	W	S	R	
DIV 1	5	0	C_5^5	$\times\ C_0^0$	$\times\ C_0^{27}$	$= 1 \times 1 \times 1 =$ 1
DIV 2	4	0	C_4^5	$\times\ C_0^0$	$\times\ C_1^{27}$	$= 5 \times 1 \times 27 =$ 135
DIV 3	3	0	C_3^5	$\times\ C_0^0$	$\times\ C_2^{27}$	$= 10 \times 1 \times 351 =$ 3,510
DIV 4	2	0	C_2^5	$\times\ C_0^0$	$\times\ C_3^{27}$	$= 10 \times 1 \times 2,925 =$ 29,250

Lotteries of this type are played in Colorado and Idaho, USA.

5/35

The lottery drawing machine contains thirty-five balls numbered from 1 to 35. Five balls are drawn without replacement.

These are the winning numbers.

Prize Structure and Probabilities

5/35

	Match Winning Numbers	Favorable Outcomes	Probabilities
			1 in
DIV 1	5	1	324,632
DIV 2	4	150	2,164
DIV 3	3	4,350	75

Total Possible Outcomes: C(35,5) = 324,632

Calculation of Favorable Outcomes

W = Winning numbers (5)   S = Suppl. number (0)   R = Remaining numbers (30)

MATCH

	W	S	W	S	R		
DIV 1	5	0	C_5^5 ×	C_0^0 ×	C_0^{30}	= 1 x 1 x 1 =	1
DIV 2	4	0	C_4^5 ×	C_0^0 ×	C_1^{30}	= 5 x 1 x 30 =	150
DIV 3	3	0	C_3^5 ×	C_0^0 ×	C_2^{30}	= 10 x 1 x 435 =	4,350

A lottery of this type is played in Connecticut, USA.

5/36

The lottery drawing machine contains thirty-six balls numbered from 1 to 36. Five balls are drawn without replacement.

These are the winning numbers.

Prize Structure and Probabilities

	5/36		
	Match		
	Winning Numbers	Favorable Outcomes	Probabilities
			1 in
DIV 1	5	1	376,992
DIV 2	4	155	2,432
DIV 3	3	4,650	81
DIV 4	2	44,950	8

Total Possible Outcomes: $C(36,5) = 376,992$

Calculation of Favorable Outcomes

W = Winning numbers (5)   S = Suppl. number (0)   R = Remaining numbers (31)

MATCH

	W	S	W	S	R		
DIV 1	5	0	C_5^5 ×	C_0^0 ×	C_0^{31}	= 1 x 1 x 1 =	1
DIV 2	4	0	C_4^5 ×	C_0^0 ×	C_1^{31}	= 5 x 1 x 31 =	155
DIV 3	3	0	C_3^5 ×	C_0^0 ×	C_2^{31}	= 10 x 1 x 465 =	4,650
DIV 4	2	0	C_2^5 ×	C_0^0 ×	C_3^{31}	= 10 x 1 x 4,495 =	44,950

A lottery of this type is played in Florida, USA.

5/39

The lottery drawing machine contains thirty-nine balls numbered from 1 to 39. Five balls are drawn without replacement.
These are the winning numbers.

Prize Structure and Probabilities

	Match Winning Numbers	Favorable Outcomes	Probabilities
5/39			
			1 in
DIV 1	5	1	575,757
DIV 2	4	170	3,387
DIV 3	3	5,610	103
DIV 4	2	59,840	10

Total Possible Outcomes: $C(39,5) = 575,757$

Calculation of Favorable Outcomes			

W = Winning numbers (5)   S = Suppl. number (0)   R = Remaining numbers (34)

MATCH

	W	S			
DIV 1	5	0	$C_5^5 \times C_0^0 \times C_0^{34}$	$= 1 \times 1 \times 1 =$	1
DIV 2	4	0	$C_4^5 \times C_0^0 \times C_1^{34}$	$= 5 \times 1 \times 34 =$	170
DIV 3	3	0	$C_3^5 \times C_0^0 \times C_2^{34}$	$= 10 \times 1 \times 561 =$	5,610
DIV 4	2	0	$C_2^5 \times C_0^0 \times C_3^{34}$	$= 10 \times 1 \times 5,984 =$	59,840

A lottery of this type is played in California, USA.

5/41

The lottery drawing machine contains forty-one balls numbered from 1 to 41. Five balls are drawn without replacement.

These are the winning numbers.

Prize Structure and Probabilities

5/41

	Match Winning Numbers	Favorable Outcomes	Probabilities
			1 in
DIV 1	5	1	749,398
DIV 2	4	180	4,163
DIV 3	3	6,300	119
DIV 4	2	71,400	10

Total Possible Outcomes: $C(41,5) = 749,398$

Calculation of Favorable Outcomes

W = Winning numbers (5)   S = Suppl. numbers (0)   R = Remaining numbers (36)

MATCH

	W	S	W	S	R		
DIV 1	5	0	C_5^5	$\times\ C_0^0$	$\times\ C_0^{36}$	$= 1 \times 1 \times 1 =$	1
DIV 2	4	0	C_4^5	$\times\ C_0^0$	$\times\ C_1^{36}$	$= 5 \times 1 \times 36 =$	180
DIV 3	3	0	C_3^5	$\times\ C_0^0$	$\times\ C_2^{36}$	$= 10 \times 1 \times 630 =$	6,300
DIV 4	2	0	C_2^5	$\times\ C_0^0$	$\times\ C_3^{36}$	$= 10 \times 1 \times 7,140 =$	71,400

A lottery of this type is played in Arizona, USA.

5/42

The lottery drawing machine contains forty-two balls numbered from 1 to 42. Five balls are drawn without replacement.

These are the winning numbers.

Prize Structure and Probabilities

5/42

	Match Winning Numbers	Favorable Outcomes	Probabilities
			1 in
DIV 1	5	1	850,668
DIV 2	4	185	4,598
DIV 3	3	6,660	128
DIV 4	2	77,700	11

Total Possible Outcomes: C(42,5) = 850,668

Calculation of Favorable Outcomes

W = Winning numbers (5)   S = Suppl. numbers (0)   R = Remaining numbers (37)

MATCH

	W	S					
DIV 1	5	0	C_5^5 ×	C_0^0 ×	C_0^{37}	= 1 x 1 x 1 =	1
DIV 2	4	0	C_4^5 ×	C_0^0 ×	C_1^{37}	= 5 x 1 x 37 =	185
DIV 3	3	0	C_3^5 ×	C_0^0 ×	C_2^{37}	= 10 x 1 x 666 =	6,660
DIV 4	2	0	C_2^5 ×	C_0^0 ×	C_3^{37}	= 10 x 1 x 7,770 =	77,700

A lottery of this type is played in Georgia, USA.

5/45

The lottery drawing machine contains forty-five balls numbered from 1 to 45. Five balls are drawn without replacement.

These are the winning numbers.

Prize Structure and Probabilities

5/45

	Match Winning Numbers	Favorable Outcomes	Probabilities
			1 in
DIV 1	5	1	1,221,759
DIV 2	4	200	6,109
DIV 3	3	7,800	157
DIV 4	2	98,800	12

Total Possible Outcomes: C(45,5) = 1,221,759

Calculation of Favorable Outcomes

W = Winning numbers (5)   **S** = Suppl. numbers (0)   R = Remaining numbers (40)

MATCH

	W	S					
DIV 1	5	0	C_5^5 × C_0^0 × C_0^{40}	= 1 x 1 x 1 =	1
DIV 2	4	0	C_4^5 × C_0^0 × C_1^{40}	= 5 x 1 x 40 =	200
DIV 3	3	0	C_3^5 × C_0^0 × C_2^{40}	= 10 x 1 x 780 =	7,800
DIV 4	2	0	C_2^5 × C_0^0 × C_3^{40}	= 10 x 1 x 9,880 =	98,800

A lottery of this type is played in Indiana, USA.

5/80

The lottery drawing machine contains eighty balls numbered from 1 to 80. Five balls are drawn without replacement.

These are the winning numbers.

Prize Structure and Probabilities

5/80

	Match Winning Numbers	Favorable Outcomes	Probabilities
			1 in
DIV 1	5	1	24,040,016
DIV 2	4	375	64,107
DIV 3	3	27,750	866
DIV 4	2	675,250	36

Total Possible Outcomes: C(80,5) = 24,040,016

Calculation of Favorable Outcomes			
W = Winning numbers (5)	S = Suppl. numbers (0)	R = Remaining numbers (75)	

MATCH

	W	S	W	S	R	
DIV 1	5	0	C_5^5 ×	C_0^0 ×	C_0^{75} = 1 x 1 x 1 =	1
DIV 2	4	0	C_4^5 ×	C_0^0 ×	C_1^{75} = 5 x 1 x 75 =	375
DIV 3	3	0	C_3^5 ×	C_0^0 ×	C_2^{75} = 10 x 1 x 2,775 =	27,750
DIV 4	2	0	C_2^5 ×	C_0^0 ×	C_3^{75} = 10 x 1 x 67,525 =	675,250

A lottery of this type is played in Brazil.

5/90

The lottery drawing machine contains ninety balls numbered from 1 to 90. Five balls are drawn without replacement.

These are the winning numbers.

Prize Structure and Probabilities

	Match Winning Numbers	Favorable Outcomes	Probabilities
			1 in
DIV 1	5	1	43,949,268
DIV 2	4	425	103,410
DIV 3	3	35,700	1,231
DIV 4	2	987,700	44

Total Possible Outcomes: C(90,5) = 43,949,268

Calculation of Favorable Outcomes

W = Winning numbers (5)   S = Suppl. numbers (0)   R = Remaining numbers (85)

MATCH

	W	S	W	S	R		
DIV 1	5	0	C_5^5 ×	C_0^0 ×	C_0^{85}	= 1 x 1 x 1 =	1
DIV 2	4	0	C_4^5 ×	C_0^0 ×	C_1^{85}	= 5 x 1 x 85 =	425
DIV 3	3	0	C_3^5 ×	C_0^0 ×	C_2^{85}	= 10 x 1 x 3,570 =	35,700
DIV 4	2	0	C_2^5 ×	C_0^0 ×	C_3^{85}	= 10 x 1 x 98,770 =	987,700

A lottery of this type is played in Hungary.

6/30 (1 supplementary number)

The lottery drawing machine contains thirty balls numbered from 1 to 30. Six balls are drawn without replacement. These are the winning numbers.
A seventh ball is drawn without replacement. This is the supplementary number.

Prize Structure and Probabilities

6/30 (1 supplementary number)				
	Match			
	Winning Numbers	Suppl. Number	Favorable Outcomes	Probabilities
				1 in
DIV 1	6		1	593,775
DIV 2	5	1	6	98,963
DIV 3	5		138	4,303
DIV 4	4	1	345	1,721
DIV 5	4		3,795	156
DIV 6	3	1	5,060	117
DIV 7	3		35,420	17

Total Possible Outcomes: $C(30,6) = 593,775$

Calculation of Favorable Outcomes

W = Winning numbers (6)   S = Suppl. numbers (1)   R = Remaining numbers (23)

MATCH

	W	S	W	S	R		
DIV 1	6	0	C_6^6 ×	C_0^1 ×	C_0^{23}	= 1 x 1 x 1 =	1
DIV 2	5	1	C_5^6 ×	C_1^1 ×	C_0^{23}	= 6 x 1 x 1 =	6
DIV 3	5	0	C_5^6 ×	C_0^1 ×	C_1^{23}	= 6 x 1 x 23 =	138
DIV 4	4	1	C_4^6 ×	C_1^1 ×	C_1^{23}	= 15 x 1 x 23 =	345
DIV 5	4	0	C_4^6 ×	C_0^1 ×	C_2^{23}	= 15 x 1 x 253 =	3,795
DIV 6	3	1	C_3^6 ×	C_1^1 ×	C_2^{23}	= 20 x 1 x 253 =	5,060
DIV 7	3	0	C_3^6 ×	C_0^1 ×	C_3^{23}	= 20 x 1 x 1,771 =	35,420

A lottery of this type is played in Lithuania.

6/40

The lottery drawing machine contains forty balls numbered from 1 to 40.
Six balls are drawn without replacement.
These are the winning numbers.

Prize Structure and Probabilities

6/40

	Match Winning Numbers	Favorable Outcomes	Probabilities
			1 in
DIV 1	6	1	3,838,380
DIV 2	5	204	18,816
DIV 3	4	8,415	456
DIV 4	3	119,680	32

Total Possible Outcomes: $C(40,6) = 3,838,380$

Calculation of Favorable Outcomes

W = Winning numbers (6)　　　S = Suppl. numbers (0)　　　R = Remaining numbers (34)

MATCH

	W	S	W	S	R		
DIV 1	6	0	C_6^6	$\times$ C_0^0	$\times$ C_0^{34}	$= 1 \times 1 \times 1 =$	1
DIV 2	5	0	C_5^6	$\times$ C_0^0	$\times$ C_1^{34}	$= 6 \times 1 \times 34 =$	204
DIV 3	4	0	C_4^6	$\times$ C_0^0	$\times$ C_2^{34}	$= 15 \times 1 \times 561 =$	8,415
DIV 4	3	0	C_3^6	$\times$ C_0^0	$\times$ C_3^{34}	$= 20 \times 1 \times 5,984 =$	119,680

A lottery of this type is played in Colorado, USA.

6/40 (1 supplementary number)

The lottery drawing machine contains forty balls numbered from 1 to 40. Six balls are drawn without replacement. These are the winning numbers.

A seventh ball is drawn without replacement. This is the supplementary number.

Prize Structure and Probabilities

6/40 (1 supplementary number)

	Match			
	Winning Numbers	Suppl. Number	Favorable Outcomes	Probabilities
				1 in
DIV 1	6		1	3,838,380
DIV 2	5	1	6	639,730
DIV 3	5		198	19,386
DIV 4	4	1	495	7,754
DIV 5	4		7,920	485
DIV 6	3	1	10,560	363
DIV 7	3		109,120	35

Total Possible Outcomes: C(40,6) = 3,838,380

Calculation of Favorable Outcomes

W = Winning numbers (6)   S = Suppl. numbers (1)   R = Remaining numbers (33)

MATCH

	W	S	W		S		R		
DIV 1	6	0	C_6^6	$\times$	C_0^1	$\times$	C_0^{33}	= 1 x 1 x 1 =	1
DIV 2	5	1	C_5^6	$\times$	C_1^1	$\times$	C_0^{33}	= 6 x 1 x 1 =	6
DIV 3	5	0	C_5^6	$\times$	C_0^1	$\times$	C_1^{33}	= 6 x 1 x 33 =	198
DIV 4	4	1	C_4^6	$\times$	C_1^1	$\times$	C_1^{33}	= 15 x 1 x 33 =	495
DIV 5	4	0	C_4^6	$\times$	C_0^1	$\times$	C_2^{33}	= 15 x 1 x 528 =	7,920
DIV 6	3	1	C_3^6	$\times$	C_1^1	$\times$	C_2^{33}	= 20 x 1 x 528 =	10,560
DIV 7	3	0	C_3^6	$\times$	C_0^1	$\times$	C_3^{33}	= 20 x 1 x 5,456 =	109,120

A lottery of this type is played in New Zealand.

6/41 (1 supplementary number)

The lottery drawing machine contains forty-one balls numbered from 1 to 41. Six balls are drawn without replacement. These are the winning numbers.

A seventh ball is drawn without replacement. This is the supplementary number.

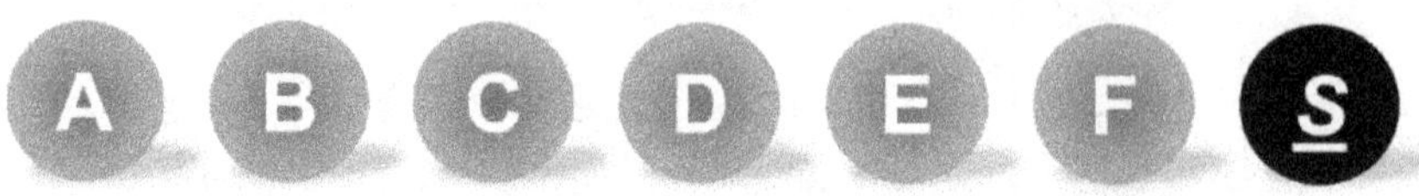

Prize Structure and Probabilities

6/40 (1 supplementary number)

| | Match | | | |
---	Winning Numbers	Suppl. Number	Favorable Outcomes	Probabilities
				1 in
DIV 1	6		1	4,496,388
DIV 2	5	1	6	749,398
DIV 3	5		204	22,041
DIV 4	4	1	510	8,816
DIV 5	4		8,415	534
DIV 6	3	1	11,220	401
DIV 7	3		119,680	38
DIV 8	2	1	89,760	50

Total Possible Outcomes: $C(41,6) = 4{,}496{,}388$

Calculation of Favorable Outcomes

W = Winning numbers (6)   **S** = Suppl. numbers (1)   **R** = Remaining numbers (34)

MATCH

	W	S	W		S		R		
DIV 1	6	0	C_6^6	x	C_0^1	x	C_0^{34}	= 1 x 1 x 1 =	1
DIV 2	5	1	C_5^6	x	C_1^1	x	C_0^{34}	= 6 x 1 x 1 =	6
DIV 3	5	0	C_5^6	x	C_0^1	x	C_1^{34}	= 6 x 1 x 34 =	204
DIV 4	4	1	C_4^6	x	C_1^1	x	C_1^{34}	= 15 x 1 x 34 =	510
DIV 5	4	0	C_4^6	x	C_0^1	x	C_2^{34}	= 15 x 1 x 561 =	8,415
DIV 6	3	1	C_3^6	x	C_1^1	x	C_2^{34}	= 20 x 1 x 561 =	11,220
DIV 7	3	0	C_3^6	x	C_0^1	x	C_3^{34}	= 20 x 1 x 5,984 =	119,680
DIV 8	2	1	C_2^6	x	C_1^1	x	C_3^{34}	= 15 x 1 x 5,984 =	89,760

A lottery of this type is played in Chile.

6/42

The lottery drawing machine contains forty-two balls numbered from 1 to 42. Six balls are drawn without replacement.

These are the winning numbers.

Prize Structure and Probabilities

	Match Winning Numbers	Favorable Outcomes	Probabilities
6/42			
			1 in
DIV 1	6	1	5,245,786
DIV 2	5	216	24,286
DIV 3	4	9,450	555
DIV 4	3	142,800	37

Total Possible Outcomes: $C(42,6) = 5,245,786$

Calculation of Favorable Outcomes				
W = Winning numbers (6)	S = Suppl. numbers (0)	R = Remaining numbers (36)		

MATCH

	W	S	W	S	R		
DIV 1	6	0	C_6^6	$\times\ C_0^0$	$\times\ C_0^{36}$	$= 1 \times 1 \times 1 =$	1
DIV 2	5	0	C_5^6	$\times\ C_0^0$	$\times\ C_1^{36}$	$= 6 \times 1 \times 36 =$	216
DIV 3	4	0	C_4^6	$\times\ C_0^0$	$\times\ C_2^{36}$	$= 15 \times 1 \times 630 =$	9,450
DIV 4	3	0	C_3^6	$\times\ C_0^0$	$\times\ C_3^{36}$	$= 20 \times 1 \times 7,140 =$	142,800

A lottery of this type is played in Arizona, USA.

6/43 (1 supplementary number)

The lottery drawing machine contains forty-three balls numbered from 1 to 43. Six balls are drawn without replacement. These are the winning numbers.

A seventh ball is drawn without replacement. This is the supplementary number.

Prize Structure and Probabilities

6/43 (1 supplementary number)

	Match			
	Winning Numbers	Suppl. Number	Favorable Outcomes	Probabilities
				1 in
DIV 1	6		1	6,096,454
DIV 2	5	1	6	1,016,076
DIV 3	5		216	28,224
DIV 4	4		9,990	610
DIV 5	3		155,400	39

Total Possible Outcomes: C(43,6) = 6,096,454

Calculation of Favorable Outcomes

W = Winning numbers (6)   **S** = Suppl. numbers (1)   **R** = Remaining numbers (36)

MATCH

	W	S	W	S	R		
DIV 1	6	0	C_6^6 ×	C_0^1 ×	C_0^{36}	= 1 x 1 x 1 =	1
DIV 2	5	1	C_5^6 ×	C_1^1 ×	C_0^{36}	= 6 x 1 x 1 =	6
DIV 3	5	0	C_5^6 ×	C_0^1 ×	C_1^{36}	= 6 x 1 x 36 =	216
DIV 4	4	1	C_4^6 ×	C_1^1 ×	C_1^{36}	= 15 x 1 x 36 = 540	9,990
	4	0	C_4^6 ×	C_0^1 ×	C_2^{36}	= 15 x 1 x 630 = 9,450	
DIV 5	3	1	C_3^6 ×	C_1^1 ×	C_2^{36}	= 20 x 1 x 630 = 12,600	155,400
	3	0	C_3^6 ×	C_0^1 ×	C_3^{36}	= 20 x 1 x 7,140 = 142,800	

A lottery of this type is played in Japan.

6/44

The lottery drawing machine contains forty-four balls numbered from 1 to 44. Six balls are drawn without replacement.
These are the winning numbers.

Prize Structure and Probabilities

6/44

	Match Winning Numbers	Favorable Outcomes	Probabilities
			1 in
DIV 1	6	1	7,059,052
DIV 2	5	228	30,961
DIV 3	4	10,545	669
DIV 4	3	168,720	42

Total Possible Outcomes: $C(44,6) = 7,059,052$

Calculation of Favorable Outcomes					

W = Winning numbers (6) S = Suppl. numbers (0) R = Remaining numbers (38)

	MATCH W	S	W	S	R		
DIV 1	6	0	C^6_6	$\times\ C^0_0$	$\times\ C^{38}_0$	$= 1 \times 1 \times 1 =$	1
DIV 2	5	0	C^6_5	$\times\ C^0_0$	$\times\ C^{38}_1$	$= 6 \times 1 \times 38 =$	228
DIV 3	4	0	C^6_4	$\times\ C^0_0$	$\times\ C^{38}_2$	$= 15 \times 1 \times 703 =$	10,545
DIV 4	3	0	C^6_3	$\times\ C^0_0$	$\times\ C^{38}_3$	$= 20 \times 1 \times 8,436 =$	168,720

Lotteries of this type are played in Arizona,
and Connecticut, USA.

6/45

The lottery drawing machine contains forty-five balls numbered from 1 to 45. Six balls are drawn without replacement.

These are the winning numbers.

Prize Structure and Probabilities

	6/45		
	Match Winning Numbers	Favorable Outcomes	Probabilities
			1 in
DIV 1	6	1	8,145,060
DIV 2	5	234	34,808
DIV 3	4	11,115	733
DIV 4	3	182,780	45

Total Possible Outcomes: C(45,6) = 8,145,060

Calculation of Favorable Outcomes					
W = Winning numbers (6)		S = Suppl. numbers (0)		R = Remaining numbers (39)	
MATCH	W S	W	S	R	
DIV 1	6 0	C_6^6 ×	C_0^0 ×	C_0^{39} = 1 x 1 x 1 =	1
DIV 2	5 0	C_5^6 ×	C_0^0 ×	C_1^{39} = 6 x 1 x 39 =	234
DIV 3	4 0	C_4^6 ×	C_0^0 ×	C_2^{39} = 15 x 1 x 741 =	11,115
DIV 4	3 0	C_3^6 ×	C_0^0 ×	C_3^{39} = 20 x 1 x 9,139 =	182,780

Lotteries of this type are played in Hungary and Philippines.

6/45 (1 supplementary number)
Prize Structure A

The lottery drawing machine contains forty-five balls numbered from 1 to 45. Six balls are drawn without replacement. These are the winning numbers.

One more ball is drawn without replacement. This is the supplementary number.

Prize Structure and Probabilities

6/45 (1 supplementary number)
Prize Structure A

	Match			
	Winning Numbers	Suppl. Number	Favorable Outcomes	Probabilities
				1 in
DIV 1	6		1	8,145,060
DIV 2	5	1	6	1,357,510
DIV 3	5		228	35,724
DIV 4	4		11,115	733
DIV 5	3		182,780	45

Total Possible Outcomes: C(45,6) = 8,145,060

Calculation of Favorable Outcomes

W = Winning numbers (6)   **S** = Suppl. numbers (1)   **R** = Remaining numbers (38)

	W	S	W	S	R		
MATCH							
DIV 1	6	0	C_6^6 ×	C_0^1 ×	C_0^{38} = 1 x 1 x 1 =		1
DIV 2	5	1	C_5^6 ×	C_1^1 ×	C_0^{38} = 6 x 1 x 1 =		6
DIV 3	5	0	C_5^6 ×	C_0^1 ×	C_1^{38} = 6 x 1 x 38 =		228
DIV 4	4	1	C_4^6 ×	C_1^1 ×	C_1^{38} = 15 x 1 x 38 = 570		11,115
	4	0	C_4^6 ×	C_0^1 ×	C_2^{38} = 15 x 1 x 703 = 10,545		
DIV 5	3	1	C_3^6 ×	C_1^1 ×	C_2^{38} = 20 x 1 x 703 = 14,060		182,780
	3	0	C_3^6 ×	C_0^1 ×	C_3^{38} = 20 x 1 x 8,436 = 168,720		

A lottery of this type is played in Croatia.

6/45 (1 supplementary number)
Prize Structure B

The lottery drawing machine contains forty-five balls numbered from 1 to 45. Six balls are drawn without replacement. These are the winning numbers.

One more ball is drawn without replacement. This is the supplementary number.

Prize Structure and Probabilities

6/45 (1 supplementary number)
Prize Structure B

	Match			
	Winning Numbers	Suppl. Number	Favorable Outcomes	Probabilities
				1 in
DIV 1	6		1	8,145,060
DIV 2	5	1	6	1,357,510
DIV 3	5		228	35,724
DIV 4	4	1	570	14,290
DIV 5	4		10,545	772
DIV 6	3	1	14,060	579
DIV 7	3		168,720	48
DIV 8	2	1	126,540	64
DIV 9	2		1,107,225	7

Total Possible Outcomes: C(45,6) = 8,145,060

Calculation of Favorable Outcomes

W = Winning numbers (6)   **S** = Suppl. number (1)   **R** = Remaining numbers (38)

MATCH

	W	S	W		S		R		
DIV 1	6	0	C_6^6	×	C_0^1	×	C_0^{38}	= 1 × 1 × 1 =	1
DIV 2	5	1	C_5^6	×	C_1^1	×	C_0^{38}	= 6 × 1 × 1 =	6
DIV 3	5	0	C_5^6	×	C_0^1	×	C_1^{38}	= 6 × 1 × 38 =	228
DIV 4	4	1	C_4^6	×	C_1^1	×	C_1^{38}	= 15 × 1 × 38 =	570
DIV 5	4	0	C_4^6	×	C_0^1	×	C_2^{38}	= 15 × 1 × 703 =	10,545
DIV 6	3	1	C_3^6	×	C_1^1	×	C_2^{38}	= 20 × 1 × 703 =	14,060
DIV 7	3	0	C_3^6	×	C_0^1	×	C_3^{38}	= 20 × 1 × 8,436 =	168,720
DIV 8	2	1	C_2^6	×	C_1^1	×	C_3^{38}	= 15 × 1 × 8,436 =	126,540
DIV 9	2	0	C_2^6	×	C_0^1	×	C_4^{38}	= 15 × 1 × 73,815 =	1,107,225

A lottery of this type is played in The Netherlands.

6/45 (1 supplementary number)
Prize Structure C

The lottery drawing machine contains forty-five balls numbered from 1 to 45. Six balls are drawn without replacement. These are the winning numbers.

One more ball is drawn without replacement. This is the supplementary number.

Prize Structure and Probabilities

6/45 (1 supplementary number)
Prize structure C

	Match			Probabilities	
	Winning Numbers	Suppl. Number	Favorable Outcomes	1 game 1 in	2 games 1 in
DIV 1	6		1	8,145,060	4,072,530
DIV 2	5	1	6	1,357,510	678,755
DIV 3	5		228	35,724	17,862
DIV 4	4	1	570	14,290	7,145
DIV 5	4		10,545	772	386
DIV 6	3	1	14,060	579	290
DIV 7	3		168,720	48	24
DIV 8		1	501,942	16	8

Total Possible Outcomes: C(45,6) = 8,145,060

Calculation of Favorable Outcomes

W = Winning numbers (6)   **S** = Suppl. numbers (1)   **R** = Remaining numbers (38)

MATCH

	W	S	W		S		R		
DIV 1	6	0	C_6^6	x	C_0^1	x	C_0^{38}	= 1 x 1 x 1 =	1
DIV 2	5	1	C_5^6	x	C_1^1	x	C_0^{38}	= 6 x 1 x 1 =	6
DIV 3	5	0	C_5^6	x	C_0^1	x	C_1^{38}	= 6 x 1 x 38 =	228
DIV 4	4	1	C_4^6	x	C_1^1	x	C_1^{38}	= 15 x 1 x 38 =	570
DIV 5	4	0	C_4^6	x	C_0^1	x	C_2^{38}	= 15 x 1 x 703 =	10,545
DIV 6	3	1	C_3^6	x	C_1^1	x	C_2^{38}	= 20 x 1 x 703 =	14,060
DIV 7	3	0	C_3^6	x	C_0^1	x	C_3^{38}	= 20 x 1 x 8,436 =	168,720
DIV 8	0	1	C_0^6	x	C_1^1	x	C_5^{38}	= 1 x 1 x 501,942 =	501,942

Lotteries of this type are played in Austria and Canada.

6/45 (1 supplementary number)
Prize Structure D

The lottery drawing machine contains forty-five balls numbered from 1 to 45. Six balls are drawn without replacement. These are the winning numbers.

One more ball is drawn without replacement. This is the supplementary number.

Prize Structure and Probabilities

6/45 (1 supplementary number)
Prize Structure D

	Match			
	Winning Numbers	Suppl. Number	Favorable Outcomes	Probabilities
				1 in
DIV 1	6		1	8,145,060
DIV 2	5	1	6	1,357,510
DIV 3	5		228	35,724
DIV 4	4	1	570	14,290
DIV 5	4		10,545	772
DIV 6	3	1	14,060	579
DIV 7	3		168,720	48
DIV 8	2	1	126,540	64
DIV 9	1	1	442,890	18

Total Possible Outcomes: C(45,6) = 8,145,060

Calculation of Favorable Outcomes

W = Winning numbers (6)   S = Suppl. numbers (1)   R = Remaining numbers (38)

MATCH

	W	S	W		S		R		
DIV 1	6	0	C_6^6	x	C_0^1	x	C_0^{38}	= 1 x 1 x 1 =	1
DIV 2	5	1	C_5^6	x	C_1^1	x	C_0^{38}	= 6 x 1 x 1 =	6
DIV 3	5	0	C_5^6	x	C_0^1	x	C_1^{38}	= 6 x 1 x 38 =	228
DIV 4	4	1	C_4^6	x	C_1^1	x	C_1^{38}	= 15 x 1 x 38 =	570
DIV 5	4	0	C_4^6	x	C_0^1	x	C_2^{38}	= 15 x 1 x 703 =	10,545
DIV 6	3	1	C_3^6	x	C_1^1	x	C_2^{38}	= 20 x 1 x 703 =	14,060
DIV 7	3	0	C_3^6	x	C_0^1	x	C_3^{38}	= 20 x 1 x 8,436 =	168,720
DIV 8	2	1	C_2^6	x	C_1^1	x	C_3^{38}	= 15 x 1 x 8,436 =	126,540
DIV 9	1	1	C_1^6	x	C_1^1	x	C_4^{38}	= 6 x 1 x 73,815 =	442,890

A lottery of this type is played in Belgium.

6/45 (2 supplementary numbers)
Prize Structure A

The lottery drawing machine contains forty-five balls numbered from 1 to 45. Six balls are drawn without replacement. These are the winning numbers.

Two more balls are drawn without replacement. These are the supplementary numbers.

Prize Structure and Probabilities

6/45 (2 supplementary numbers)
Prize Structure A

	Match Winning Numbers	Suppl. Numbers	Favorable Outcomes	Probabilities
				1 in
DIV 1	6		1	8,145,060
DIV 2	5	1	12	678,755
DIV 3	5		222	36,689
DIV 4	4		11,115	733
DIV 5	3	1	27,380	297
DIV 6	3		155,400	52

Total Possible Outcomes: C(45,6) = 8,145,060

Calculation of Favorable Outcomes

W = Winning numbers (6)   **S** = Suppl. numbers (2)   R = Remaining numbers (37)

MATCH

	W	S	W		S		R			
DIV 1	6	0	C_6^6	×	C_0^2	×	C_0^{37}	= 1 x 1 x 1 =		1
DIV 2	5	1	C_5^6	×	C_1^2	×	C_0^{37}	= 6 x 2 x 1 =		12
DIV 3	5	0	C_5^6	×	C_0^2	×	C_1^{37}	= 6 x 1 x 37 =		222
DIV 4	4	0	C_4^6	×	C_0^2	×	C_2^{37}	= 15 x 1 x 666 =	9,990	
	4	1	C_4^6	×	C_1^2	×	C_1^{37}	= 15 x 2 x 37 =	1,110	11,115
	4	2	C_4^6	×	C_2^2	×	C_0^{37}	= 15 x 1 x 1 =	15	
DIV 5	3	1	C_3^6	×	C_1^2	×	C_2^{37}	= 20 x 2 x 666 =	26,640	
	3	2	C_3^6	×	C_2^2	×	C_1^{37}	= 20 x 1 x 37 =	740	27,380
DIV 6	3	0	C_3^6	×	C_0^2	×	C_3^{37}	= 20 x 1 x 7,770 =		155,400

A lottery of this type is played in Australia.

6/45 (2 supplementary numbers)
Prize Structure B

The lottery drawing machine contains forty-five balls numbered from 1 to 45. Six balls are drawn without replacement. These are the winning numbers.

Two more balls are drawn without replacement. These are the supplementary numbers.

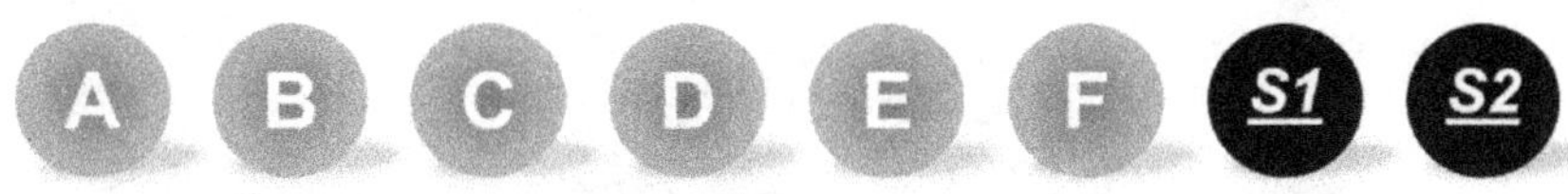

Prize Structure and Probabilities

6/45 (2 supplementary numbers)
Prize Structure B

	Match			
	Winning Numbers	Suppl. Numbers	Favorable Outcomes	Probabilities
				1 in
DIV 1	6		1	8,145,060
DIV 2	5	1	12	678,755
DIV 3	5		222	36,689
DIV 4	4		11,115	733
DIV 5	3	1	27,380	297
DIV 6	1 or 2	2	56,610	144

Total Possible Outcomes: $C(45,6) = 8,145,060$

Calculation of Favorable Outcomes

W = Winning numbers (6)   S = Suppl. numbers (2)   R = Remaining numbers (37)

MATCH

	W	S	W		S		R			
DIV 1	6	0	C_6^6	×	C_0^2	×	C_0^{37}	= 1 x 1 x 1 =		1
DIV 2	5	1	C_5^6	×	C_1^2	×	C_0^{37}	= 6 x 2 x 1 =		12
DIV 3	5	0	C_5^6	×	C_0^2	×	C_1^{37}	= 6 x 1 x 37 =		222
DIV 4	4	0	C_4^6	×	C_0^2	×	C_2^{37}	= 15 x 1 x 666 =	9,990	
	4	1	C_4^6	×	C_1^2	×	C_1^{37}	= 15 x 2 x 37 =	1,110	11,115
	4	2	C_4^6	×	C_2^2	×	C_0^{37}	= 15 x 1 x 1 =	15	
DIV 5	3	1	C_3^6	×	C_1^2	×	C_2^{37}	= 20 x 2 x 666 =	26,640	
	3	2	C_3^6	×	C_2^2	×	C_1^{37}	= 20 x 1 x 37 =	740	27,380
DIV 6	1	2	C_1^6	×	C_2^2	×	C_3^{37}	= 6 x 1 x 7,770 =	46,620	
	2	2	C_2^6	×	C_2^2	×	C_2^{37}	= 15 x 1 x 666 =	9,990	56,610

A lottery of this type is played in Australia.

6/46

The lottery drawing machine contains forty-six balls numbered from 1 to 46. Six balls are drawn without replacement. These are the winning numbers.

Prize Structure and Probabilities

6/46

	Match Winning Numbers	Favorable Outcomes	Probabilities
			1 in
DIV 1	6	1	9,366,819
DIV 2	5	240	39,028
DIV 3	4	11,700	801
DIV 4	3	197,600	47
DIV 5	2	1,370,850	7

Total Possible Outcomes: $C(46,6) = 9,366,819$

Calculation of Favorable Outcomes

W = Winning numbers (6)   S = Suppl. numbers (0)   R = Remaining numbers (40)

MATCH

	W	S	W	S	R		
DIV 1	6	0	C_6^6	$\times C_0^0 \times$	C_0^{40}	$= 1 \times 1 \times 1 =$	1
DIV 2	5	0	C_5^6	$\times C_0^0 \times$	C_1^{40}	$= 6 \times 1 \times 40 =$	240
DIV 3	4	0	C_4^6	$\times C_0^0 \times$	C_2^{40}	$= 15 \times 1 \times 780 =$	11,700
DIV 4	3	0	C_3^6	$\times C_0^0 \times$	C_3^{40}	$= 20 \times 1 \times 9,880 =$	197,600
DIV 5	2	0	C_2^6	$\times C_0^0 \times$	C_4^{40}	$= 15 \times 1 \times 91,390 =$	1,370,850

A lottery of this type is played in Indiana, USA.

6/47

The lottery drawing machine contains forty-seven balls numbered from 1 to 47. Six balls are drawn without replacement.

These are the winning numbers.

Prize Structure and Probabilities

6/47

	Match Winning Numbers	Favorable Outcomes	Probabilities
			1 in
DIV 1	6	1	10,737,573
DIV 2	5	246	43,649
DIV 3	4	12,300	873
DIV 4	3	213,200	50

Total Possible Outcomes: $C(47,6) = 10,737,573$

Calculation of Favorable Outcomes

W = Winning numbers (6)   S = Suppl. numbers (0)   R = Remaining numbers (41)

MATCH

	W	S	W	S	R		
DIV 1	6	0	C_6^6 ×	C_0^0 ×	C_0^{41}	$= 1 \times 1 \times 1 =$	1
DIV 2	5	0	C_5^6 ×	C_0^0 ×	C_1^{41}	$= 6 \times 1 \times 41 =$	246
DIV 3	4	0	C_4^6 ×	C_0^0 ×	C_2^{41}	$= 15 \times 1 \times 820 =$	12,300
DIV 4	3	0	C_3^6 ×	C_0^0 ×	C_3^{41}	$= 20 \times 1 \times 10,660 =$	213,200

A lottery of this type is played in Georgia, USA.

6/47 (1 supplementary number)

The lottery drawing machine contains forty-seven balls numbered from 1 to 47. Six balls are drawn without replacement. These are the winning numbers.

One more ball is drawn without replacement. This is the supplementary number.

Prize Structure and Probabilities

6/47 (1 supplementary number)
Prize Structure B

	Match Winning Numbers	Suppl. Number	Favorable Outcomes	Probabilities
				1 in
DIV 1	6		1	10,737,573
DIV 2	5	1	6	1,789,596
DIV 3	5		240	44,740
DIV 4	4	1	600	17,896
DIV 5	4		11,700	918
DIV 6	3	1	15,600	688
DIV 7	3		197,600	54
DIV 8	2	1	148,200	72

Total Possible Outcomes: C(47,6) = 10,737,573

Calculation of Favorable Outcomes

W = Winning numbers (6)   S = Suppl. numbers (1)   R = Remaining numbers (40)

	MATCH		W	S	R		
	W	**S**	**W**	**S**	**R**		
DIV 1	6	0	C_6^6	$\times$ C_0^1	$\times$ C_0^{40}	= 1 x 1 x 1 =	1
DIV 2	5	1	C_5^6	$\times$ C_1^1	$\times$ C_0^{40}	= 6 x 1 x 1 =	6
DIV 3	5	0	C_5^6	$\times$ C_0^1	$\times$ C_1^{40}	= 6 x 1 x 40 =	240
DIV 4	4	1	C_4^6	$\times$ C_1^1	$\times$ C_1^{40}	= 15 x 1 x 40 =	600
DIV 5	4	0	C_4^6	$\times$ C_0^1	$\times$ C_2^{40}	= 15 x 1 x 780 =	11,700
DIV 6	3	1	C_3^6	$\times$ C_1^1	$\times$ C_2^{40}	= 20 x 1 x 780 =	15,600
DIV 7	3	0	C_3^6	$\times$ C_0^1	$\times$ C_3^{40}	= 20 x 1 x 9,880 =	197,600
DIV 8	2	1	C_2^6	$\times$ C_1^1	$\times$ C_3^{40}	= 15 x 1 x 9,880 =	148,200

A lottery of this type is played in Ireland.

6/49

The lottery drawing machine contains forty-nine balls numbered from 1 to 49. Six balls are drawn without replacement.

These are the winning numbers.

Prize Structure and Probabilities

6/49

	Match Winning Numbers	Favorable Outcomes	Probabilities
			1 in
DIV 1	6	1	13,983,816
DIV 2	5	258	54,201
DIV 3	4	13,545	1,032
DIV 4	3	246,820	57

Total Possible Outcomes: $C(49,6) = 13,983,816$

Calculation of Favorable Outcomes						

W = Winning numbers (6) S = Suppl. numbers (0) R = Remaining numbers (43)

MATCH

	W	S	W	S	R		
DIV 1	6	0	C_6^6	$\times\ C_0^0$	$\times\ C_0^{43}$	$= 1 \times 1 \times 1 =$	1
DIV 2	5	0	C_5^6	$\times\ C_0^0$	$\times\ C_1^{43}$	$= 6 \times 1 \times 43 =$	258
DIV 3	4	0	C_4^6	$\times\ C_0^0$	$\times\ C_2^{43}$	$= 15 \times 1 \times 903 =$	13,545
DIV 4	3	0	C_3^6	$\times\ C_0^0$	$\times\ C_3^{43}$	$= 20 \times 1 \times 12,341 =$	246,820

Lotteries of this type are played in Philippines and Poland.

6/49 (1 supplementary number)
Prize Structure A

The lottery drawing machine contains forty-nine balls numbered from 1 to 49. Six balls are drawn without replacement. These are the winning numbers.

One more ball is drawn without replacement. This is the supplementary number.

Prize Structure and Probabilities

6/49 (1 supplementary number)
Prize Structure A

	Match			
	Winning Numbers	Suppl. Number	Favorable Outcomes	Probabilities
				1 in
DIV 1	6		1	13,983,816
DIV 2	5	1	6	2,330,636
DIV 3	5		252	55,491
DIV 4	4		13,545	1,032
DIV 5	3		246,820	57

Total Possible Outcomes: C(49,6) = 13,983,816

Calculation of Favorable Outcomes

W = Winning numbers (6)   **S** = Suppl. numbers (1)   R = Remaining numbers (42)

MATCH

	W	S	W		S		R		
DIV 1	6	0	C_6^6	×	C_0^1	×	C_0^{42}	$= 1 \times 1 \times 1 =$	1
DIV 2	5	1	C_5^6	×	C_1^1	×	C_0^{42}	$= 6 \times 1 \times 1 =$	6
DIV 3	5	0	C_5^6	×	C_0^1	×	C_1^{42}	$= 6 \times 1 \times 42 =$	252
DIV 4	4	1	C_4^6	×	C_1^1	×	C_1^{42}	$= 15 \times 1 \times 42 = 630$	13,545
	4	0	C_4^6	×	C_0^1	×	C_2^{42}	$= 15 \times 1 \times 861 = 12,915$	
DIV 5	3	1	C_3^6	×	C_1^1	×	C_2^{42}	$= 20 \times 1 \times 861 = 17,220$	246,820
	3	0	C_3^6	×	C_0^1	×	C_3^{42}	$= 20 \times 1 \times 11,480 = 229,600$	

A lottery of this type is played in Greece.

6/49 (1 supplementary number)
Prize Structure B

The lottery drawing machine contains forty-nine balls numbered from 1 to 49. Six balls are drawn without replacement. These are the winning numbers.

One more ball is drawn without replacement. This is the supplementary number.

Prize Structure and Probabilities

6/49 (1 supplementary number)
Prize Structure B

	Match			
	Winning Numbers	Suppl. Number	Favorable Outcomes	Probabilities
				1 in
DIV 1	6		1	13,983,816
DIV 2	5	1	6	2,330,636
DIV 3	5		252	55,491
DIV 4	4		13,545	1,032
DIV 5	3		246,820	57
DIV 6	2	1	172,200	81
DIV 7	2		1,678,950	8

Total Possible Outcomes: C(49,6) = 13,983,816

Calculation of Favorable Outcomes

W = Winning numbers (6)   **S** = Suppl. numbers (1)   **R** = Remaining numbers (42)

MATCH

	W	S	W	S	R		
DIV 1	6	0	C_6^6 ×	C_0^1 ×	C_0^{42}	= 1 x 1 x 1 =	1
DIV 2	5	1	C_5^6 ×	C_1^1 ×	C_0^{42}	= 6 x 1 x 1 =	6
DIV 3	5	0	C_5^6 ×	C_0^1 ×	C_1^{42}	= 6 x 1 x 42 =	252
DIV 4	4	1	C_4^6 ×	C_1^1 ×	C_1^{42}	= 15 x 1 x 42 =	630 ⎤
	4	0	C_4^6 ×	C_0^1 ×	C_2^{42}	= 15 x 1 x 861 =	12,915 ⎦ 13,545
DIV 5	3	1	C_3^6 ×	C_1^1 ×	C_2^{42}	= 20 x 1 x 861 =	17,220 ⎤
	3	0	C_3^6 ×	C_0^1 ×	C_3^{42}	= 20 x 1 x 11,480 =	229,600 ⎦ 246,820
DIV 6	2	1	C_2^6 ×	C_1^1 ×	C_3^{42}	= 15 x 1 x 11,480 =	172,200
DIV 7	2	0	C_2^6 ×	C_0^1 ×	C_4^{42}	= 15 x 1 x 111,930 =	1,678,950

A lottery of this type is played in Canada.

6/49 (1 supplementary number)
Prize Structure C

The lottery drawing machine contains forty-nine balls numbered from 1 to 49. Six balls are drawn without replacement. These are the winning numbers.

One more ball is drawn without replacement. This is the supplementary number.

Prize Structure and Probabilities

6/49 (1 supplementary number)
Prize Structure C

	Match Winning Numbers	Suppl. Number	Favorable Outcomes	Probabilities
				1 in
DIV 1	6		1	13,983,816
DIV 2	5	1	6	2,330,636
DIV 3	5		252	55,491
DIV 4	4	1	630	22,197
DIV 5	4		12,915	1,083
DIV 6	3	1	17,220	812
DIV 7	3		229,600	61

Total Possible Outcomes: C(49,6) = 13,983,816

Calculation of Favorable Outcomes

W = Winning numbers (6)   S = Suppl. numbers (1)   R = Remaining numbers (42)

MATCH

	W	S	W	S	R		
DIV 1	6	0	C_6^6	$\times$ C_0^1	$\times$ C_0^{42}	= 1 x 1 x 1 =	1
DIV 2	5	1	C_5^6	$\times$ C_1^1	$\times$ C_0^{42}	= 6 x 1 x 1 =	6
DIV 3	5	0	C_5^6	$\times$ C_0^1	$\times$ C_1^{42}	= 6 x 1 x 42 =	252
DIV 4	4	1	C_4^6	$\times$ C_1^1	$\times$ C_1^{42}	= 15 x 1 x 42 =	630
DIV 5	4	0	C_4^6	$\times$ C_0^1	$\times$ C_2^{42}	= 15 x 1 x 861 =	12,915
DIV 6	3	1	C_3^6	$\times$ C_1^1	$\times$ C_2^{42}	= 20 x 1 x 861 =	17,220
DIV 8	3	0	C_3^6	$\times$ C_0^1	$\times$ C_3^{42}	= 20 x 1 x 11,480 =	229,600

Lotteries of this type are played in Hong Kong and Singapore.

6/49 (1 supplementary number)
Prize Structure D

The lottery drawing machine contains forty-nine balls numbered from 1 to 49. Six balls are drawn without replacement. These are the winning numbers.

One more ball is drawn without replacement. This is the supplementary number.

Prize Structure and Probabilities

6/49 (1 supplementary number)
Prize Structure D

	Match			
	Winning Numbers	Suppl. Number	Favorable Outcomes	Probabilities
				1 in
DIV 1	6		1	13,983,816
DIV 2	5	1	6	2,330,636
DIV 3	5		252	55,491
DIV 4	4		13,545	1,032
DIV 5	3	1	17,220	812
DIV 6	3		229,600	61
DIV 7	2	1	172,200	81

Total Possible Outcomes: C(49,6) = 13,983,816

Calculation of Favorable Outcomes

W = Winning numbers (6)   S = Suppl. numbers (1)   R = Remaining numbers (42)

MATCH	W	S	W		S		R			
DIV 1	6	0	C^6_6	x	C^1_0	x	C^{42}_0	= 1 x 1 x 1 =		1
DIV 2	5	1	C^6_5	x	C^1_1	x	C^{42}_0	= 6 x 1 x 1 =		6
DIV 3	5	0	C^6_5	x	C^1_0	x	C^{42}_1	= 6 x 1 x 42 =		252
DIV 4	4	1	C^6_4	x	C^1_1	x	C^{42}_1	= 15 x 1 x 42 =	630	13,545
	4	0	C^6_4	x	C^1_0	x	C^{42}_2	= 15 x 1 x 861 =	12,915	
DIV 5	3	1	C^6_3	x	C^1_1	x	C^{42}_2	= 20 x 1 x 861 =		17,220
DIV 6	3	0	C^6_3	x	C^1_0	x	C^{42}_3	= 20 x 1 x 11,480 =		229,600
DIV 7	2	1	C^6_2	x	C^1_1	x	C^{42}_3	= 15 x 1 x 11,480 =		172,200

A lottery of this type is played in Slovakia.

6/52

The lottery drawing machine contains fifty-two balls numbered from 1 to 52. Six balls are drawn without replacement. These are the winning numbers.

Prize Structure and Probabilities

	6/52		
	Match		
	Winning Numbers	Favorable Outcomes	Probabilities
			1 in
DIV 1	6	1	20,358,520
DIV 2	5	276	73,763
DIV 3	4	15,525	1,311
DIV 4	3	303,600	67
DIV 5	2	2,447,775	8

Total Possible Outcomes: $C(52,6) = 20,358,520$

Calculation of Favorable Outcomes

W = Winning numbers (6)   S = Suppl. numbers (0)   R = Remaining numbers (46)

MATCH

	W	S	W	S	R		
DIV 1	6	0	C_6^6	$\times\ C_0^0$	$\times\ C_0^{46}$	$= 1 \times 1 \times 1 =$	1
DIV 2	5	0	C_5^6	$\times\ C_0^0$	$\times\ C_1^{46}$	$= 6 \times 1 \times 46 =$	276
DIV 3	4	0	C_4^6	$\times\ C_0^0$	$\times\ C_2^{46}$	$= 15 \times 1 \times 1,035 =$	15,525
DIV 4	3	0	C_3^6	$\times\ C_0^0$	$\times\ C_3^{46}$	$= 20 \times 1 \times 15,180 =$	303,600
DIV 5	2	0	C_2^6	$\times\ C_0^0$	$\times\ C_4^{46}$	$= 15 \times 1 \times 163,185 =$	2,447,775

A lottery of this type is played in Ukraine.

6/53

The lottery drawing machine contains fifty-three balls numbered from 1 to 53. Six balls are drawn without replacement. These are the winning numbers.

Prize Structure and Probabilities

	Match Winning Numbers	Favorable Outcomes	Probabilities
6/53			
			1 in
DIV 1	6	1	22,957,480
DIV 2	5	282	81,410
DIV 3	4	16,215	1,416
DIV 4	3	324,300	71
DIV 5	2	2,675,475	9

Total Possible Outcomes: $C(53,6) = 22,957,480$

Calculation of Favorable Outcomes				
W = Winning numbers (6)	**S** = Suppl. numbers (0)	**R** = Remaining numbers (47)		

	MATCH W / S	W	S	R	
DIV 1	6 / 0	C^6_6	$\times\ C^0_0$	$\times\ C^{47}_0$	$= 1 \times 1 \times 1 =$ 1
DIV 2	5 / 0	C^6_5	$\times\ C^0_0$	$\times\ C^{47}_1$	$= 6 \times 1 \times 47 =$ 282
DIV 3	4 / 0	C^6_4	$\times\ C^0_0$	$\times\ C^{47}_2$	$= 15 \times 1 \times 1,081 =$ 16,215
DIV 4	3 / 0	C^6_3	$\times\ C^0_0$	$\times\ C^{47}_3$	$= 20 \times 1 \times 16,215 =$ 324,300
DIV 5	2 / 0	C^6_2	$\times\ C^0_0$	$\times\ C^{47}_4$	$= 15 \times 1 \times 178,365 =$ 2,675,475

A lottery of this type is played in Florida, USA.

6/54

The lottery drawing machine contains fifty-four balls numbered from 1 to 54. Six balls are drawn without replacement.

These are the winning numbers.

Prize Structure and Probabilities

	6/54		
	Match Winning Numbers	Favorable Outcomes	Probabilities
			1 in
DIV 1	6	1	25,827,165
DIV 2	5	288	89,678
DIV 3	4	16,920	1,526
DIV 4	3	345,920	75

Total Possible Outcomes: C(54,6) = 25,827,165

Calculation of Favorable Outcomes					
W = Winning numbers (6)		S = Suppl. numbers (0)		R = Remaining numbers (48)	

W = Winning numbers (6) S = Suppl. numbers (0) R = Remaining numbers (48)

MATCH

	W	S	W	S	R		
DIV 1	6	0	C_6^6	$\times\ C_0^0$	$\times\ C_0^{48}$	= 1 x 1 x 1 =	1
DIV 2	5	0	C_5^6	$\times\ C_0^0$	$\times\ C_1^{48}$	= 6 x 1 x 48 =	288
DIV 3	4	0	C_4^6	$\times\ C_0^0$	$\times\ C_2^{48}$	= 15 x 1 x 1,128 =	16,920
DIV 4	3	0	C_3^6	$\times\ C_0^0$	$\times\ C_3^{48}$	= 20 x 1 x 17,296 =	345,920

A lottery of this type is played in Texas, USA.

6/55

The lottery drawing machine contains fifty-five balls numbered from 1 to 55. Six balls are drawn without replacement.

These are the winning numbers.

Prize Structure and Probabilities

6/55

	Match Winning Numbers	Favorable Outcomes	Probabilities
			1 in
DIV 1	6	1	28,989,675
DIV 2	5	294	98,604
DIV 3	4	17,640	1,643
DIV 4	3	368,480	79

Total Possible Outcomes: C(55,6) = 28,989,675

Calculation of Favorable Outcomes

W = Winning numbers (6)   S = Suppl. numbers (0)   R = Remaining numbers (49)

MATCH

	W	S	W	S	R		
DIV 1	6	0	C_6^6	C_0^0	C_0^{49}	= 1 x 1 x 1 =	1
DIV 2	5	0	C_5^6	C_0^0	C_1^{49}	= 6 x 1 x 49 =	294
DIV 3	4	0	C_4^6	C_0^0	C_2^{49}	= 15 x 1 x 1,176 =	17,640
DIV 4	3	0	C_3^6	C_0^0	C_3^{49}	= 20 x 1 x 18,424 =	368,480

A lottery of this type is played in Philippines.

6/56 (1 supplementary number)

The lottery ball drawing machine contains fifty-six balls numbered from 1 to 56. Six balls are drawn without replacement. These are the winning numbers.

One more ball is drawn without replacement. This is the supplementary number.

Prize Structure and Probabilities

6/56 (1 supplementary number)

	Match			
	Winning Numbers	Suppl. Number	Favorable Outcomes	Probabilities
				1 in
DIV 1	6		1	32,468,436
DIV 2	5	1	6	5,411,406
DIV 3	5		294	110,437
DIV 4	4	1	735	44,175
DIV 5	4		17,640	1,841
DIV 6	3	1	23,520	1,380
DIV 7	3		368,480	88
DIV 8	2	1	276,360	117
DIV 9	2		3,178,140	10

Total Possible Outcomes: C(56,6) = 32,468,436

Calculation of Favorable Outcomes

W = Winning numbers (6)   **S** = Suppl. numbers (1)   **R** = Remaining numbers (49)

MATCH

	W	S	W		S		R		
DIV 1	6	0	C_6^6	x	C_0^1	x	C_0^{49}	= 1 x 1 x 1 =	1
DIV 2	5	1	C_5^6	x	C_1^1	x	C_0^{49}	= 6 x 1 x 1 =	6
DIV 3	5	0	C_5^6	x	C_0^1	x	C_1^{49}	= 6 x 1 x 49 =	294
DIV 4	4	1	C_4^6	x	C_1^1	x	C_1^{49}	= 15 x 1 x 49 =	735
DIV 5	4	0	C_4^6	x	C_0^1	x	C_2^{49}	= 15 x 1 x 1,176 =	17,640
DIV 6	3	1	C_3^6	x	C_1^1	x	C_2^{49}	= 20 x 1 x 1,176 =	23,520
DIV 7	3	0	C_3^6	x	C_0^1	x	C_3^{49}	= 20 x 1 x 18,424 =	368,480
DIV 8	2	1	C_2^6	x	C_1^1	x	C_3^{49}	= 15 x 1 x 18,424 =	276,360
DIV 9	2	0	C_2^6	x	C_0^1	x	C_4^{49}	= 15 x 1 x 211,876 =	3,178,140

A lottery of this type is played in Mexico.

6/58

The lottery drawing machine contains fifty-eight balls numbered from 1 to 58. Six balls are drawn without replacement.

These are the winning numbers.

Prize Structure and Probabilities

6/58			
	Match Winning Numbers	Favorable Outcomes	Probabilities
			1 in
DIV 1	6	1	40,475,358
DIV 2	5	312	129,729
DIV 3	4	19,890	2,035
DIV 4	3	442,000	92

Total Possible Outcomes: $C(58,6) = 40,475,358$

Calculation of Favorable Outcomes

W = Winning numbers (6)   S = Suppl. numbers (0)   R = Remaining numbers (52)

MATCH

	W	S	W	S	R		
DIV 1	6	0	C_6^6	$\times\ C_0^0$	$\times\ C_0^{52}$	$= 1 \times 1 \times 1 =$	1
DIV 2	5	0	C_5^6	$\times\ C_0^0$	$\times\ C_1^{52}$	$= 6 \times 1 \times 52 =$	312
DIV 3	4	0	C_4^6	$\times\ C_0^0$	$\times\ C_2^{52}$	$= 15 \times 1 \times 1,326 =$	19,890
DIV 4	3	0	C_3^6	$\times\ C_0^0$	$\times\ C_3^{52}$	$= 20 \times 1 \times 22,100 =$	442,000

A lottery of this type is played in Philippines.

6/59 (1 supplementary number)
Prize Structure A

The lottery drawing machine contains fifty-nine balls numbered from 1 to 59. Six balls are drawn without replacement. These are the winning numbers.

One more ball is drawn without replacement. This is the supplementary number.

Prize Structure and Probabilities

6/59 (1 supplementary number)
Prize Structure A

	Match			
	Winning Numbers	Suppl. Number	Favorable Outcomes	Probabilities
				1 in
DIV 1	6		1	45,057,474
DIV 2	5	1	6	7,509,579
DIV 3	5		312	144,415
DIV 4	4		20,670	2,180
DIV 5	3		468,520	96

Total Possible Outcomes: C(59,6) = 45,057,474

Calculation of Favorable Outcomes

W = Winning numbers (6)   S = Suppl. numbers (1)   R = Remaining numbers (52)

MATCH

	W	S	W		S		R		
DIV 1	6	0	C_6^6	$\times$	C_0^1	$\times$	C_0^{52}	$= 1 \times 1 \times 1 =$	1
DIV 2	5	1	C_5^6	$\times$	C_1^1	$\times$	C_0^{52}	$= 6 \times 1 \times 1 =$	6
DIV 3	5	0	C_5^6	$\times$	C_0^1	$\times$	C_1^{52}	$= 6 \times 1 \times 52 =$	312
DIV 4	4	1	C_4^6	$\times$	C_1^1	$\times$	C_1^{52}	$= 15 \times 1 \times 52 = 780$	20,670
	4	0	C_4^6	$\times$	C_0^1	$\times$	C_2^{52}	$= 15 \times 1 \times 1,326 = 19,890$	
DIV 5	3	1	C_3^6	$\times$	C_1^1	$\times$	C_2^{52}	$= 20 \times 1 \times 1,326 = 26,520$	468,520
	3	0	C_3^6	$\times$	C_0^1	$\times$	C_3^{52}	$= 20 \times 1 \times 22,100 = 442,000$	

A lottery of this type is played in New York, USA.

6/59 (1 supplementary number)
Prize Structure B

The lottery drawing machine contains fifty-nine balls numbered from 1 to 59. Six balls are drawn without replacement. These are the winning numbers.

One more ball is drawn without replacement. This is the supplementary number.

Prize Structure and Probabilities

6/59 (1 supplementary number)
Prize Structure B

| | Match | | | |
	Winning Numbers	Suppl. Number	Favorable Outcomes	Probabilities
				1 in
DIV 1	6		1	45,057,474
DIV 2	5	1	6	7,509,579
DIV 3	5		312	144,415
DIV 4	4		20,670	2,180
DIV 5	3		468,520	96
DIV 6	2		4,392,375	10

Total Possible Outcomes: C(59,6) = 45,057,474

Calculation of Favorable Outcomes

W = Winning numbers (6)   **S** = Suppl. numbers (1)   **R** = Remaining numbers (52)

MATCH					
	W	**S**	**W**	**S**	**R**	
DIV 1	6	0	C_6^6 × C_0^1 × C_0^{52}	= 1 × 1 × 1 =	1
DIV 2	5	1	C_5^6 × C_1^1 × C_0^{52}	= 6 × 1 × 1 =	6
DIV 3	5	0	C_5^6 × C_0^1 × C_1^{52}	= 6 × 1 × 52 =	312
DIV 4	4	1	C_4^6 × C_1^1 × C_1^{52}	= 15 × 1 × 52 = 780	20,670
	4	0	C_4^6 × C_0^1 × C_2^{52}	= 15 × 1 × 1,326 = 19,890	
DIV 5	3	1	C_3^6 × C_1^1 × C_2^{52}	= 20 × 1 × 1,326 = 26,520	468,520
	3	0	C_3^6 × C_0^1 × C_3^{52}	= 20 × 1 × 22,100 = 442,000	
DIV 6	2	1	C_2^6 × C_1^1 × C_3^{52}	= 15 × 1 × 22,100 = 331,500	4,392,375
	2	0	C_2^6 × C_0^1 × C_4^{52}	= 15 × 1 × 270,725 = 4,060,875	

A lottery of this type is played in the United Kingdom.

6/60

The lottery drawing machine contains sixty balls numbered from 1 to 60. Six balls are drawn without replacement.

These are the winning numbers.

Prize Structure and Probabilities

6/60

	Match Winning Numbers	Favorable Outcomes	Probabilities 1 in
DIV 1	6	1	50,063,860
DIV 2	5	324	154,518
DIV 3	4	21,465	2,332

Total Possible Outcomes: $C(60,6) = 50,063,860$

Calculation of Favorable Outcomes

W = Winning numbers (6)   S = Suppl. numbers (0)   R = Remaining numbers (54)

MATCH

	W	S	W	S	R		
DIV 1	6	0	C_6^6 ×	C_0^0 ×	C_0^{54}	= 1 × 1 × 1 =	1
DIV 2	5	0	C_5^6 ×	C_0^0 ×	C_1^{54}	= 6 × 1 × 54 =	324
DIV 3	4	0	C_4^6 ×	C_0^0 ×	C_2^{54}	= 15 × 1 × 1,431 =	21,465

A lottery of this type is played in Brazil.

6/90 (1 supplementary number)

The lottery drawing machine contains ninety balls numbered from 1 to 90. Six balls are drawn without replacement. These are the winning numbers.

One more ball is drawn without replacement. This is the supplementary number.

Prize Structure and Probabilities

6/90 (1 supplementary number)

	Match			
	Winning Numbers	Suppl. Number	Favorable Outcomes	Probabilities
				1 in
DIV 1	6		1	622,614,630
DIV 2	5	1	6	103,769,105
DIV 3	5		498	1,250,230
DIV 4	4		52,290	11,907
DIV 5	3		1,905,680	327
DIV 6	2		28,942,515	22

Total Possible Outcomes: C(90,6) = 622,614,630

Calculation of Favorable Outcomes

W = Winning numbers (6)   **S** = Suppl. numbers (1)   **R** = Remaining numbers (83)

MATCH

	W	S	W		S		R			
DIV 1	6	0	C_6^6	×	C_0^1	×	C_0^{83}	= 1 x 1 x 1 =		1
DIV 2	5	1	C_5^6	×	C_1^1	×	C_0^{83}	= 6 x 1 x 1 =		6
DIV 3	5	0	C_5^6	×	C_0^1	×	C_1^{83}	= 6 x 1 x 83 =		498
DIV 4	4	1	C_4^6	×	C_1^1	×	C_1^{83}	= 15 x 1 x 83 = 1,245		52,290
	4	0	C_4^6	×	C_0^1	×	C_2^{83}	= 15 x 1 x 3,403 = 51,045		
DIV 5	3	1	C_3^6	×	C_1^1	×	C_2^{83}	= 20 x 1 x 3,403 = 68,060		1,905,680
	3	0	C_3^6	×	C_0^1	×	C_3^{83}	= 20 x 1 x 91,881 = 1,837,620		
DIV 6	2	1	C_2^6	×	C_1^1	×	C_3^{83}	= 15 x 1 x 91,881 = 1,378,215		28,942,515
	2	0	C_2^6	×	C_0^1	×	C_4^{83}	= 15 x 1 x 1,837,620 = 27,564,300		

A lottery of this type is played in Italy.

7/34 (1 supplementary number)

The lottery drawing machine contains thirty-four balls numbered from 1 to 34. Seven balls are drawn without replacement. These are the winning numbers.

One more ball is drawn without replacement. This is the supplementary number.

Prize Structure and Probabilities

7/34 (1 supplementary number)

	Match			
	Winning Numbers	Suppl. Number	Favorable Outcomes	Probabilities
				1 in
DIV 1	7		1	5,379,616
DIV 2	6	1	7	768,517
DIV 3	6		182	29,558
DIV 4	5		7,371	730
DIV 5	4		102,375	53

Total Possible Outcomes: C(34,7) = 5,379,616

Calculation of Favorable Outcomes

W = Winning numbers (7)   **S** = Suppl. numbers (1)   **R** = Remaining numbers (26)

MATCH

	W	S	**W**		**S**		**R**		
DIV 1	7	0	C_7^7	×	C_0^1	×	C_0^{26}	= 1 x 1 x 1 =	1
DIV 2	6	1	C_6^7	×	C_1^1	×	C_0^{26}	= 7 x 1 x 1 =	7
DIV 3	6	0	C_6^7	×	C_0^1	×	C_1^{26}	= 7 x 1 x 26 =	182
DIV 4	5	1	C_5^7	×	C_1^1	×	C_1^{26}	= 21 x 1 x 26 = 546	7,371
	5	0	C_5^7	×	C_0^1	×	C_2^{26}	= 21 x 1 x 325 = 6,825	
DIV 5	4	1	C_4^7	×	C_1^1	×	C_2^{26}	= 35 x 1 x 325 = 11,375	102,375
	4	0	C_4^7	×	C_0^1	×	C_3^{26}	= 35 x 1 x 2,600 = 91,000	

A lottery of this type is played in Norway.

7/35 (1 supplementary number)

The lottery drawing machine contains thirty-five balls numbered from 1 to 35. Seven balls are drawn without replacement. These are the winning numbers.

One more ball is drawn without replacement. This is the supplementary number.

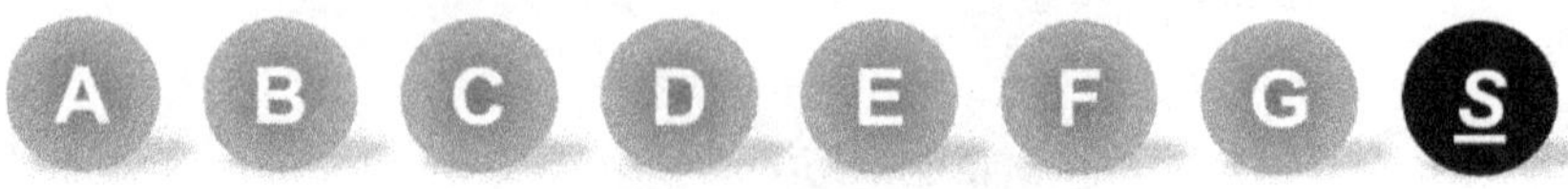

Prize Structure and Probabilities

7/35 (1 supplementary number)

	Match			
	Winning Numbers	Suppl. Number	Favorable Outcomes	Probabilities
				1 in
DIV 1	7		1	6,724,520
DIV 2	6	1	7	960,646
DIV 3	6		189	35,579
DIV 4	5	1	567	11,860
DIV 5	5		7,371	912
DIV 6	4	1	12,285	547
DIV 7	4		102,375	66
DIV 8	3	1	102,375	66
DIV 9		1	296,010	23

Total Possible Outcomes: $C(35,7) = 6{,}724{,}520$

Calculation of Favorable Outcomes

W = Winning numbers (7)   **S** = Suppl. numbers (1)   R = Remaining numbers (27)

MATCH

	W	S	W		S		R	
DIV 1	7	0	C_7^7	x	C_0^1	x	C_0^{27}	= 1 x 1 x 1 = 1
DIV 2	6	1	C_6^7	x	C_1^1	x	C_0^{27}	= 7 x 1 x 1 = 7
DIV 3	6	0	C_6^7	x	C_0^1	x	C_1^{27}	= 7 x 1 x 27 = 189
DIV 4	5	1	C_5^7	x	C_1^1	x	C_1^{27}	= 21 x 1 x 27 = 567
DIV 5	5	0	C_5^7	x	C_0^1	x	C_2^{27}	= 21 x 1 x 351 = 7,371
DIV 6	4	1	C_4^7	x	C_1^1	x	C_2^{27}	= 35 x 1 x 351 = 12,285
DIV 7	4	0	C_4^7	x	C_0^1	x	C_3^{27}	= 35 x 1 x 2,925 = 102,375
DIV 8	3	1	C_3^7	x	C_1^1	x	C_3^{27}	= 35 x 1 x 2,925 = 102,375
DIV 9	0	1	C_0^7	x	C_1^1	x	C_6^{27}	= 1 x 1 x 296,010 = 296,010

A lottery of this type is played in Croatia.

7/35 (4 supplementary numbers)

The lottery ball drawing machine contains thirty-five balls numbered from 1 to 35. Seven balls are drawn without replacement. These are the winning numbers.

Four more balls are drawn without replacement. These are the supplementary numbers.

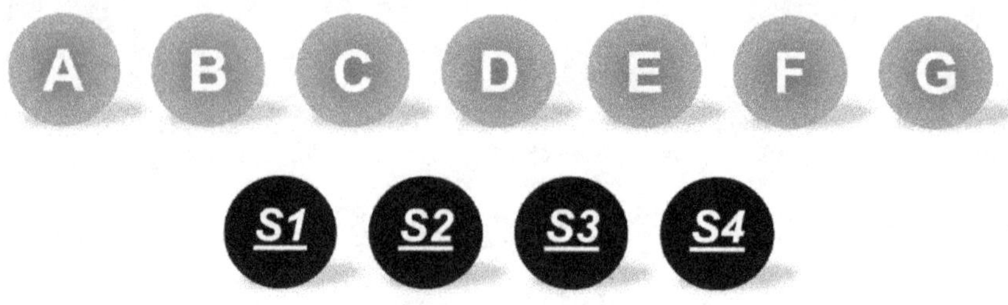

Prize Structure and Probabilities

	7/35 (4 supplementary numbers)			
	Match			
	Winning Numbers	Suppl. Numbers	Favorable Outcomes	Probabilities
				1 in
DIV 1	7		1	6,724,520
DIV 2	6	1	28	240,161
DIV 3	6		168	40,027
DIV 4	5		7,938	847
DIV 5	4		114,660	59
DIV 6	3		716,625	9

Total Possible Outcomes: $C(35,7) = 6,724,520$

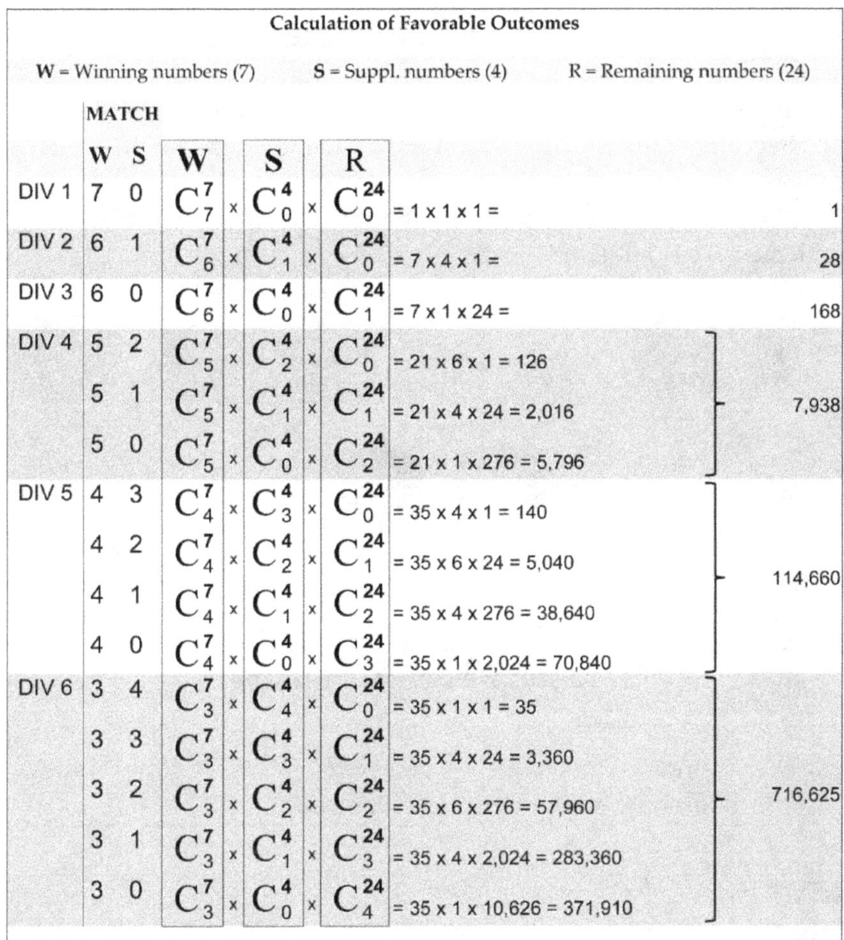

Calculation of Favorable Outcomes

W = Winning numbers (7)   **S** = Suppl. numbers (4)   **R** = Remaining numbers (24)

	MATCH					
	W	**S**	**W**	**S**	**R**	
DIV 1	7	0	C_7^7 ×	C_0^4 ×	C_0^{24}	= 1 x 1 x 1 = ... 1
DIV 2	6	1	C_6^7 ×	C_1^4 ×	C_0^{24}	= 7 x 4 x 1 = ... 28
DIV 3	6	0	C_6^7 ×	C_0^4 ×	C_1^{24}	= 7 x 1 x 24 = ... 168
DIV 4	5	2	C_5^7 ×	C_2^4 ×	C_0^{24}	= 21 x 6 x 1 = 126
	5	1	C_5^7 ×	C_1^4 ×	C_1^{24}	= 21 x 4 x 24 = 2,016
	5	0	C_5^7 ×	C_0^4 ×	C_2^{24}	= 21 x 1 x 276 = 5,796
						7,938
DIV 5	4	3	C_4^7 ×	C_3^4 ×	C_0^{24}	= 35 x 4 x 1 = 140
	4	2	C_4^7 ×	C_2^4 ×	C_1^{24}	= 35 x 6 x 24 = 5,040
	4	1	C_4^7 ×	C_1^4 ×	C_2^{24}	= 35 x 4 x 276 = 38,640
	4	0	C_4^7 ×	C_0^4 ×	C_3^{24}	= 35 x 1 x 2,024 = 70,840
						114,660
DIV 6	3	4	C_3^7 ×	C_4^4 ×	C_0^{24}	= 35 x 1 x 1 = 35
	3	3	C_3^7 ×	C_3^4 ×	C_1^{24}	= 35 x 4 x 24 = 3,360
	3	2	C_3^7 ×	C_2^4 ×	C_2^{24}	= 35 x 6 x 276 = 57,960
	3	1	C_3^7 ×	C_1^4 ×	C_3^{24}	= 35 x 4 x 2,024 = 283,360
	3	0	C_3^7 ×	C_0^4 ×	C_4^{24}	= 35 x 1 x 10,626 = 371,910
						716,625

A lottery of this type is played in Sweden.

7/37 (2 supplementary numbers)

The lottery drawing machine contains thirty-seven balls numbered from 1 to 37. Seven balls are drawn without replacement. These are the winning numbers.

Two more balls are drawn without replacement. These are the supplementary numbers.

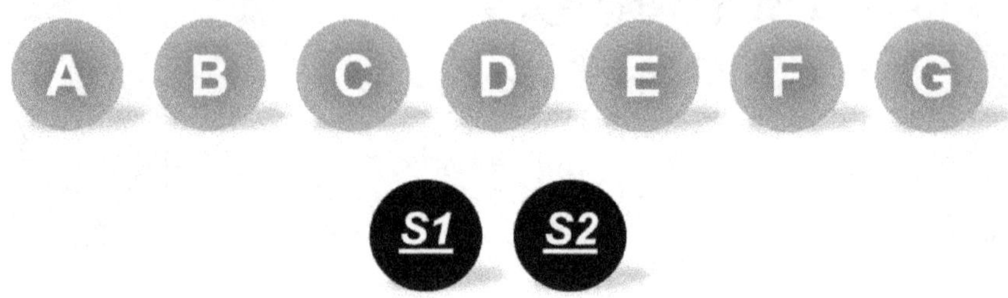

Prize Structure and Probabilities

7/37 (2 supplementary numbers)

	Match		Favorable Outcomes	Probabilities
	Winning Numbers	Suppl. Numbers		
				1 in
DIV 1	7		1	10,295,472
DIV 2	6	1	14	735,391
DIV 3	6		196	52,528
DIV 4	5		9,135	1,127
DIV 5	4		142,100	72
DIV 6	3	1	242,550	42

Total Possible Outcomes: C(37,7) = 10,295,472

Calculation of Favorable Outcomes

W = Winning numbers (7)   **S** = Suppl. numbers (2)   **R** = Remaining numbers (28)

MATCH

	W	S	W	S	R		
DIV 1	7	0	C_7^7 ×	C_0^2 ×	C_0^{28}	= 1 x 1 x 1 =	1
DIV 2	6	1	C_6^7 ×	C_1^2 ×	C_0^{28}	= 7 x 2 x 1 =	14
DIV 3	6	0	C_6^7 ×	C_0^2 ×	C_1^{28}	= 7 x 1 x 28 =	196
DIV 4	5	2	C_5^7 ×	C_2^2 ×	C_0^{28}	= 21 x 1 x 1 =	21
	5	1	C_5^7 ×	C_1^2 ×	C_1^{28}	= 21 x 2 x 28 =	1,176
	5	0	C_5^7 ×	C_0^2 ×	C_2^{28}	= 21 x 1 x 378 =	7,938
							9,135
DIV 5	4	2	C_4^7 ×	C_2^2 ×	C_1^{28}	= 35 x 1 x 28 =	980
	4	1	C_4^7 ×	C_1^2 ×	C_2^{28}	= 35 x 2 x 378 =	26,460
	4	0	C_4^7 ×	C_0^2 ×	C_3^{28}	= 35 x 1 x 3,276 =	114,660
							142,100
DIV 6	3	2	C_3^7 ×	C_2^2 ×	C_2^{28}	= 35 x 1 x 378 =	13,230
	3	1	C_3^7 ×	C_1^2 ×	C_3^{28}	= 35 x 2 x 3,276 =	229,320
							242,550

A lottery of this type is played in Japan.

7/40 (1 supplementary number)

The lottery drawing machine contains forty balls numbered from 1 to 40. Seven balls are drawn without replacement. These are the winning numbers.

One more ball is drawn without replacement. This is the supplementary number.

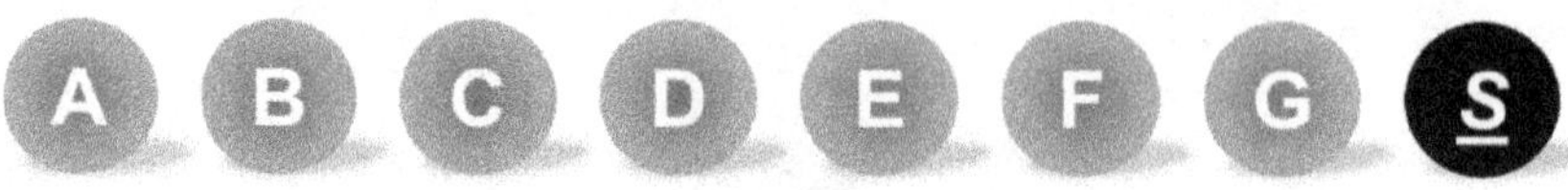

Prize Structure and Probabilities

7/40 (1 supplementary number)

	Match			
	Winning Numbers	Suppl. Number	Favorable Outcomes	Probabilities
				1 in
DIV 1	7		1	18,643,560
DIV 2	6	1	7	2,663,366
DIV 3	6		224	83,230
DIV 4	5		11,088	1,681
DIV 5	4		190,960	98
DIV 6	3	1	173,600	107

Total Possible Outcomes: C(40,7) = 18,643,560

Calculation of Favorable Outcomes

W = Winning numbers (7)   **S** = Suppl. numbers (1)   **R** = Remaining numbers (32)

MATCH

	W	S	W	S	R		
DIV 1	7	0	C^7_7 ×	C^1_0 ×	C^{32}_0	= 1 x 1 x 1 =	1
DIV 2	6	1	C^7_6 ×	C^1_1 ×	C^{32}_0	= 7 x 1 x 1 =	7
DIV 3	6	0	C^7_6 ×	C^1_0 ×	C^{32}_1	= 7 x 1 x 32 =	224
DIV 4	5	1	C^7_5 ×	C^1_1 ×	C^{32}_1	= 21 x 1 x 32 = 672	
	5	0	C^7_5 ×	C^1_0 ×	C^{32}_2	= 21 x 1 x 496 = 10,416	11,088
DIV 5	4	1	C^7_4 ×	C^1_1 ×	C^{32}_2	= 35 x 1 x 496 = 17,360	
	4	0	C^7_4 ×	C^1_0 ×	C^{32}_3	= 35 x 1 x 4,960 = 173,600	190,960
DIV 6	3	1	C^7_3 ×	C^1_1 ×	C^{32}_3	= 35 x 1 x 4,960 = 173,600	173,600
	3	0	C^7_3 ×	C^1_0 ×	C^{32}_4	= 35 x 1 x 35,960 = 1,258,600	

A lottery of this type is played in Finland.

7/44 (2 supplementary numbers)

The lottery drawing machine contains forty-four balls numbered from 1 to 44. Seven balls are drawn without replacement. These are the winning numbers.

Two more balls are drawn without replacement. These are the supplementary numbers.

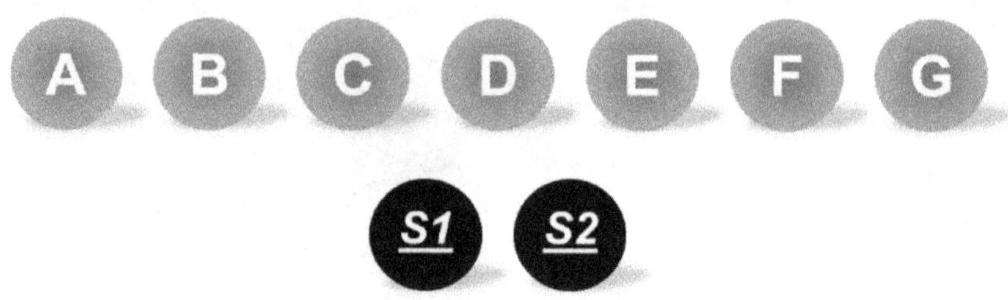

Prize Structure and Probabilities

7/44 (2 supplementary numbers)

	Match			
	Winning Numbers	Suppl. Numbers	Favorable Outcomes	Probabilities
				1 in
DIV 1	7		1	38,320,568
DIV 2	6	1	14	2,737,183
DIV 3	6		245	156,410
DIV 4	5	1	1,491	25,701
DIV 5	5		12,495	3,067
DIV 6	4	1	42,875	894
DIV 7	4		229,075	167
DIV 8	3	1	478,975	80

Total Possible Outcomes: $C(44,7) = 38,320,568$

Calculation of Favorable Outcomes

W = Winning numbers (7)   **S** = Suppl. numbers (2)   **R** = Remaining numbers (35)

MATCH

	W	S	W	S	R			
DIV 1	7	0	C_7^7	$\times\ C_0^2$	$\times\ C_0^{35}$	$= 1 \times 1 \times 1 =$		1
DIV 2	6	1	C_6^7	$\times\ C_1^2$	$\times\ C_0^{35}$	$= 7 \times 2 \times 1 =$		14
DIV 3	6	0	C_6^7	$\times\ C_0^2$	$\times\ C_1^{35}$	$= 7 \times 1 \times 35 =$		245
DIV 4	5	1	C_5^7	$\times\ C_1^2$	$\times\ C_1^{35}$	$= 21 \times 2 \times 35 =$	1,470	1,491
	5	2	C_5^7	$\times\ C_2^2$	$\times\ C_0^{35}$	$= 21 \times 1 \times 1 =$	21	
DIV 5	5	0	C_5^7	$\times\ C_0^2$	$\times\ C_2^{35}$	$= 21 \times 1 \times 595 =$		12,495
DIV 6	4	1	C_4^7	$\times\ C_1^2$	$\times\ C_2^{35}$	$= 35 \times 2 \times 595 =$	41,650	42,875
	4	2	C_4^7	$\times\ C_2^2$	$\times\ C_1^{35}$	$= 35 \times 1 \times 35 =$	1,225	
DIV 7	4	0	C_4^7	$\times\ C_0^2$	$\times\ C_3^{35}$	$= 35 \times 1 \times 6,545 =$		229,075
DIV 8	3	1	C_3^7	$\times\ C_1^2$	$\times\ C_3^{35}$	$= 35 \times 2 \times 6,545 =$	458,150	478,975
	3	2	C_3^7	$\times\ C_2^2$	$\times\ C_2^{35}$	$= 35 \times 1 \times 595 =$	20,825	

A lottery of this type is played in Australia.

7/45 (2 supplementary numbers)

The lottery drawing machine contains forty-five balls numbered from 1 to 45. Seven balls are drawn without replacement. These are the winning numbers.

Two more balls are drawn without replacement. These are the supplementary numbers.

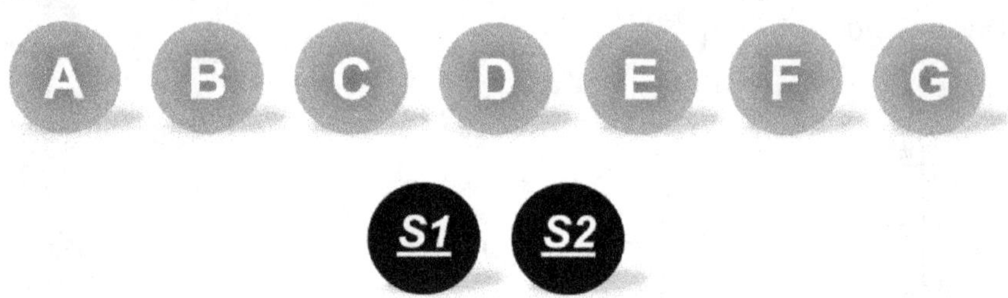

Prize Structure and Probabilities

	7/45 (2 supplementary numbers)			
	Match			
	Winning Numbers	Suppl. Numbers	Favorable Outcomes	Probabilities
				1 in
DIV 1	7		1	45,379,620
DIV 2	6	1	14	3,241,401
DIV 3	6		252	180,078
DIV 4	5	1	1,533	29,602
DIV 5	5		13,230	3,430
DIV 6	4		295,260	154
DIV 7	3	1	521,850	87

Total Possible Outcomes: $C(45,7) = 45,379,620$

Calculation of Favorable Outcomes

W = Winning numbers (7)   **S** = Suppl. numbers (2)   **R** = Remaining numbers (36)

MATCH

	W	S	W		S		R			
DIV 1	7	0	C^7_7	×	C^2_0	×	C^{36}_0	= 1 x 1 x 1 =		1
DIV 2	6	1	C^7_6	×	C^2_1	×	C^{36}_0	= 7 x 2 x 1 =		14
DIV 3	6	0	C^7_6	×	C^2_0	×	C^{36}_1	= 7 x 1 x 36 =		252
DIV 4	5	1	C^7_5	×	C^2_1	×	C^{36}_1	= 21 x 2 x 36 =	1,512	
	5	2	C^7_5	×	C^2_2	×	C^{36}_0	= 21 x 1 x 1 =	21	1,533
DIV 5	5	0	C^7_5	×	C^2_0	×	C^{36}_2	= 21 x 1 x 630 =		13,230
DIV 6	4	0	C^7_4	×	C^2_0	×	C^{36}_3	= 35 x 1 x 7,140 =	249,900	
	4	1	C^7_4	×	C^2_1	×	C^{36}_2	= 35 x 2 x 630 =	44,100	295,260
	4	2	C^7_4	×	C^2_2	×	C^{36}_1	= 35 x 1 x 36 =	1,260	
DIV 7	3	1	C^7_3	×	C^2_1	×	C^{36}_3	= 35 x 2 x 7,140 =	499,800	
	3	2	C^7_3	×	C^2_2	×	C^{36}_2	= 35 x 1 x 630 =	22,050	521,850

A lottery of this type was played in Australia
until 17 May 2022.

7/47 (3 supplementary numbers)

The lottery drawing machine contains forty-seven balls numbered from 1 to 47. Seven balls are drawn without replacement. These are the winning numbers.

Three more balls are drawn without replacement. These are the supplementary numbers.

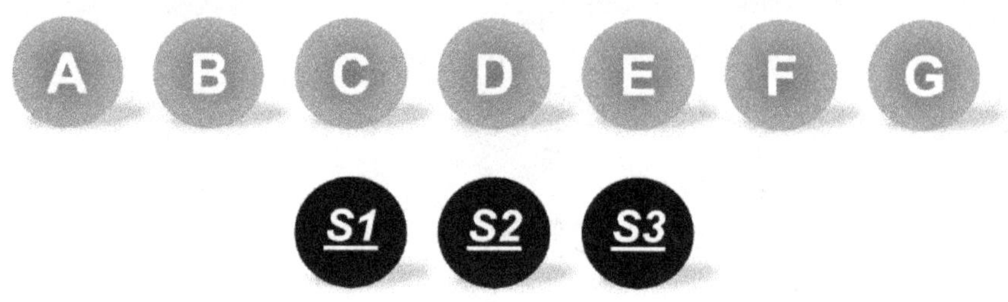

Prize Structure and Probabilities

7/47 (2 supplementary numbers)

	Match			
	Winning Numbers	Suppl. Numbers	Favorable Outcomes	Probabilities
				1 in
DIV 1	7		1	62,891,499
DIV 2	6	1	21	2,994,833
DIV 3	6		259	242,824
DIV 4	5	1	2,394	26,270
DIV 5	5		13,986	4,497
DIV 6	4		345,800	182
DIV 7	3	1	887,075	71

Total Possible Outcomes: $C(47,7) = 62,891,499$

Calculation of Favorable Outcomes

W = Winning numbers (7)   S = Suppl. numbers (3)   R = Remaining numbers (37)

	MATCH		W		S		R		
	W	S							
DIV 1	7	0	C_7^7	×	C_0^3	×	C_0^{37}	$= 1 \times 1 \times 1 =$	1
DIV 2	6	1	C_6^7	×	C_1^3	×	C_0^{37}	$= 7 \times 3 \times 1 =$	21
DIV 3	6	0	C_6^7	×	C_0^3	×	C_1^{37}	$= 7 \times 1 \times 37 =$	259
DIV 4	5	2	C_5^7	×	C_2^3	×	C_0^{37}	$= 21 \times 3 \times 1 = 63$	
	5	1	C_5^7	×	C_1^3	×	C_1^{37}	$= 21 \times 3 \times 37 = 2{,}331$	2,394
DIV 5	5	0	C_5^7	×	C_0^3	×	C_2^{37}	$= 21 \times 1 \times 666 = 13{,}986$	13,986
DIV 6	4	3	C_4^7	×	C_3^3	×	C_0^{37}	$= 35 \times 1 \times 1 = 35$	
	4	2	C_4^7	×	C_2^3	×	C_1^{37}	$= 35 \times 3 \times 37 = 3{,}885$	
	4	1	C_4^7	×	C_1^3	×	C_2^{37}	$= 35 \times 3 \times 666 = 69{,}930$	345,800
	4	0	C_4^7	×	C_0^3	×	C_3^{37}	$= 35 \times 1 \times 7{,}770 = 271{,}950$	
DIV 7	3	3	C_3^7	×	C_3^3	×	C_1^{37}	$= 35 \times 1 \times 37 = 1{,}295$	
	3	2	C_3^7	×	C_2^3	×	C_2^{37}	$= 35 \times 3 \times 666 = 69{,}930$	887,075
	3	1	C_3^7	×	C_1^3	×	C_3^{37}	$= 35 \times 3 \times 7{,}770 = 815{,}850$	

A lottery of this type is played in Australia.

7/50 (1 supplementary number)

A lottery drawing machine contains fifty balls numbered from 1 to 50. Seven balls are drawn without replacement. These are the winning numbers.

One more ball is drawn without replacement. This is the supplementary number.

Prize Structure and Probabilities

7/50 (1 supplementary number)					
	Match				
	Winning Numbers	Suppl. Number	Favorable Outcomes	Probabilities	
				1 game 1 in	3 games 1 in
DIV 1	7		1	99,884,400	33,294,800
DIV 2	6	1	7	14,269,200	4,756,400
DIV 3	6		294	339,743	113,248
DIV 4	5	1	882	113,248	37,749
DIV 5	5		18,081	5,524	1,841
DIV 6	4	1	30,135	3,315	1,105
DIV 7	4		401,800	249	83
DIV 8	3	1	401,800	249	83
DIV 9	3		3,917,550	25	8

Total Possible Outcomes: C(50,7) = 99,884,400

Calculation of Favorable Outcomes

W = Winning numbers (7)   **S** = Suppl. numbers (1)   **R** = Remaining numbers (42)

MATCH

	W	S	W		S		R	
DIV 1	7	0	C_7^7	$\times$	C_0^1	$\times$	C_0^{42}	= 1 x 1 x 1 = 1
DIV 2	6	1	C_6^7	$\times$	C_1^1	$\times$	C_0^{42}	= 7 x 1 x 1 = 7
DIV 3	6	0	C_6^7	$\times$	C_0^1	$\times$	C_1^{42}	= 7 x 1 x 42 = 294
DIV 4	5	1	C_5^7	$\times$	C_1^1	$\times$	C_1^{42}	= 21 x 1 x 42 = 882
DIV 5	5	0	C_5^7	$\times$	C_0^1	$\times$	C_2^{42}	= 21 x 1 x 861 = 18,081
DIV 6	4	1	C_4^7	$\times$	C_1^1	$\times$	C_2^{42}	= 35 x 1 x 861 = 30,135
DIV 7	4	0	C_4^7	$\times$	C_0^1	$\times$	C_3^{42}	= 35 x 1 x 11,480 = 401,800
DIV 8	3	1	C_3^7	$\times$	C_1^1	$\times$	C_3^{42}	= 35 x 1 x 11,480 = 401,800
DIV 9	3	0	C_3^7	$\times$	C_0^1	$\times$	C_4^{42}	= 35 x 1 x 111,930 = 3,917,550

A lottery of this type is played in Canada.

11/22

The lottery drawing machine contains twenty-two balls numbered from 1 to 22. Eleven balls are drawn without replacement.

These are the winning numbers.

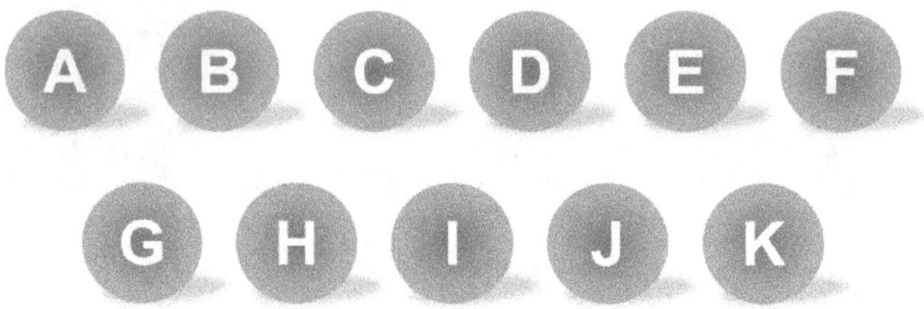

Prize Structure and Probabilities

	Match Winning Numbers	Favorable Outcomes	Probabilities
			1 in
DIV 1	11	1	705,432
DIV 2	10	121	5,830
DIV 3	9	3,025	233
DIV 4	8	27,225	26
DIV 5	3	27,225	26
DIV 6	2	3,025	233
DIV 7	1	121	5,830
DIV 8	0	1	705,432

Total Possible Outcomes: C(22,11) = 705,432

Calculation of Favorable Outcomes

W = Winning numbers (11)   **S** = Suppl. numbers (0)   **R** = Remaining numbers (11)

MATCH

	W	S	W		S		R			
DIV 1	11	0	C_{11}^{11}	×	C_{0}^{0}	×	C_{0}^{11}	= 1 x 1 x 1 =	1
DIV 2	10	0	C_{10}^{11}	×	C_{0}^{0}	×	C_{1}^{11}	= 11 x 1 x 11 =	121
DIV 3	9	0	C_{9}^{11}	×	C_{0}^{0}	×	C_{2}^{11}	= 55 x 1 x 55 =	3,025
DIV 4	8	0	C_{8}^{11}	×	C_{0}^{0}	×	C_{3}^{11}	= 165 x 1 x 165 =	27,225
DIV 5	3	0	C_{3}^{11}	×	C_{0}^{0}	×	C_{8}^{11}	= 165 x 1 x 165 =	27,225
DIV 6	2	0	C_{2}^{11}	×	C_{0}^{0}	×	C_{9}^{11}	= 55 x 1 x 55 =	3,025
DIV 7	1	0	C_{1}^{11}	×	C_{0}^{0}	×	C_{10}^{11}	= 11 x 1 x 11 =	121
DIV 8	0	0	C_{0}^{11}	×	C_{0}^{0}	×	C_{11}^{11}	= 1 x 1 x 1 =	1

A lottery of this type is played in Wisconsin, USA.

20/80

The lottery ball drawing machine contains eighty balls numbered from 1 to 80. Twenty balls are drawn without replacement. These are the winning numbers.

The player can select from one number up to twenty numbers on the betting ticket. Depending on the numbers selected there will be up to 20 different prize structures.

Total Possible Outcomes

The balls numbered from 1 to 20 represent the twenty winning numbers:

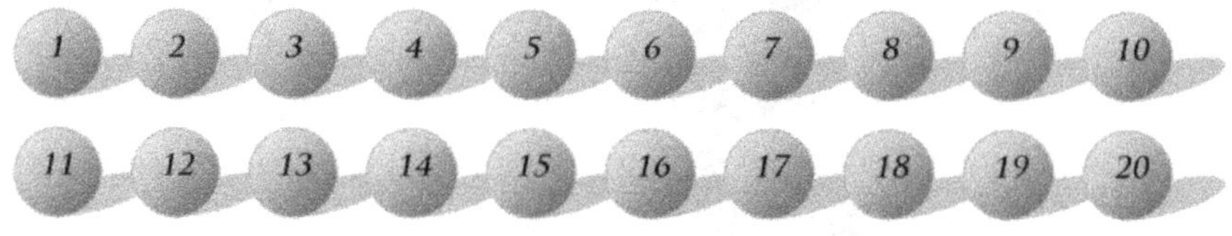

WINNING NUMBERS

Because order is not important in this lottery, the number of combinations without replacement when drawing 20 out of 80 numbers is

$$C(80,20) = 3{,}535{,}316{,}142{,}212{,}170{,}000$$

NOTE:

In this particular case Excel© calculation is wrong. The result obtained is

$$C(80,20) = 3{,}535{,}316{,}142{,}212{,}180{,}000$$

Online calculators show different values:

$$C(80,20) = 3{,}535{,}316{,}142{,}212{,}174{,}000$$
$$C(80,20) = 3{,}535{,}316{,}142{,}212{,}174{,}320$$

The best way to confirm the actual result is by simplifying the fraction, as follows:

1. Arrange both numerator and denominator in ascending order

$$\frac{61}{1} \times \frac{62}{2} \times \frac{63}{3} \times \frac{64}{4} \times \frac{65}{5} \times \frac{66}{6} \times \frac{67}{7} \times \frac{68}{8} \times \frac{69}{9} \times \frac{70}{10} \times \frac{71}{11} \times \frac{72}{12} \times \frac{73}{13} \times \frac{74}{14} \times \frac{75}{15} \times \frac{76}{16} \times \frac{77}{17} \times \frac{78}{18} \times \frac{79}{19} \times \frac{80}{20}$$

2. Do the following simplifications:

$$\frac{80}{20} = \frac{4}{1} \quad \frac{76}{19} = \frac{4}{1} \quad \frac{72}{18} = \frac{4}{1} \quad \frac{68}{17} = \frac{4}{1} \quad \frac{64}{16} = \frac{4}{1} \quad \frac{75}{15} = \frac{5}{1} \quad \frac{70}{14} = \frac{5}{1} \quad \frac{65}{13} = \frac{5}{1} \quad \frac{77}{11} = \frac{7}{1} \quad \frac{63}{9} = \frac{7}{1}$$

$$\frac{66}{6} = \frac{11}{1} \quad \frac{69}{3} = \frac{23}{1} \quad \frac{62}{2} = \frac{31}{1}$$

3. The new fraction is

$$\frac{4}{1} \times \frac{4}{1} \times \frac{4}{1} \times \frac{4}{1} \times \frac{4}{1} \times \frac{5}{1} \times \frac{5}{1} \times \frac{5}{1} \times \frac{7}{1} \times \frac{7}{1} \times \frac{11}{1} \times \frac{23}{1} \times \frac{31}{1} \times \frac{79}{12} \times \frac{78}{10} \times \frac{74}{8} \times \frac{73}{7} \times \frac{71}{5} \times \frac{67}{4} \times \frac{61}{1}$$

$$= \frac{475,146,489,513,316,000,000,000}{134,400} = 3,535,316,142,212,170,000$$

There are 3,535,316,142,212,170,000 total possible outcomes for C(80,20).

Favorable Outcomes and Probabilities

There is a general formula to calculate the probabilities in this lottery:

$$P(\text{Match } k) = C(20, k) \times C(60, n - k)/C(80, n)$$

Where:

k = total of numbers matched by the player.

n = number of selections made by the player.

Some examples are given below:

20/80 — One number selected

Match 0		
	k = 0	n − k = 1	n = 1
P(Match k) =	$C(20, k) \times C(60, n - k)/C(80, n)$		
P(Match 0) =	$C(20,0) \times C(60,1)/C(80,1)$		
	$1 \times 60/80 = 0.75$		
Probability of winning Match 0 1 in 80/60 = 1 in 1.3			
Match 1		
	k = 1	n − k = 0	n = 1
P(Match k) =	$C(20, k) \times C(60, n - k)/C(80, n)$		
P(Match 1) =	$C(20,1) \times C(60,0)/C(80,1)$		
	$20 \times 1/80 = 0.25$		
Probability of winning Match 1 1 in 80/20 = 1 in 4			

20/80 — Two numbers selected

Match 0		
	k = 0	n − k = 2	n = 2
P(Match k) =	C(20, k) × C(60, n − k)/C(80, n)		
P(Match 0) =	C(20,0) × C(60,2)/C(80,2)		
	1 × 1,770/3,160 = 0.56		
Probability of winning Match 0		
1 in 3,160/1,770 = 1 in 1.8		
Match 1		
	k = 1	n − k = 1	n = 2
P(Match k) =	C(20, k) × C(60, n − k)/C(80, n)		
P(Match 1) =	C(20,1) × C(60,1)/C(80,2)		
	20 × 60/3,160 = 0.38		
Probability of winning Match 1		
1 in 3,160/1,200 = 1 in 2.6		
Match 2		
	k = 2	n − k = 0	n = 2
P(Match k) =	C(20, k) × C(60, n − k)/C(80, n)		
P(Match 2) =	C(20,2) × C(60,0)/C(80,2)		
	190 × 1/3,160 = 0.06		
Probability of winning Match 2		
1 in 3,160/190 = 1 in 16.6		

For all other 20/80 prize structures only the results are shown in the following tables.

20/80 — One number selected

	Favorable Outcomes	Probabilities
		1 in
Match 0:	60	1.3
Match 1:	20	4

	Winning numbers		Other numbers							Favorable outcomes	Probability of winning 1 in
MATCH 0	$_{20}C_0$	×	$_{60}C_1$	=	1	×	60	=		60	1.3
MATCH 1	$_{20}C_1$	×	$_{60}C_0$	=	20	×	1	=		20	4.0
						Possible outcomes $_{80}C_1$		=		80	

20/80 — Two numbers selected

	Favorable Outcomes	Probabilities
		1 in
Match 0:	1,770	1.8
Match 1:	1,200	2.6
Match 2:	190	16.6

	Winning numbers		Other numbers							Favorable outcomes	Probability of winning 1 in
MATCH 0	$_{20}C_0$	×	$_{80}C_2$	=	1	×	1,770	=		1,770	1.8
MATCH 1	$_{20}C_1$	×	$_{60}C_1$	=	20	×	60	=		1,200	2.6
MATCH 2	$_{20}C_2$	×	$_{60}C_0$	=	190	×	1	=		190	16.6
						Possible outcomes $_{80}C_2$		=		3,160	

20/80 — Three numbers selected

	Favorable Outcomes	Probabilities
		1 in
Match 0:	34,220	2.4
Match 1:	35,400	2.3
Match 2:	11,400	7.2
Match 3:	1,140	72

	Winning numbers		Other numbers						Favorable outcomes	Probability of winning
										1 in
MATCH 0	$_{20}C_0$	×	$_{60}C_3$	=	1	×	34,220	=	34,220	2.4
MATCH 1	$_{20}C_1$	×	$_{60}C_2$	=	20	×	1,770	=	35,400	2.3
MATCH 2	$_{20}C_2$	×	$_{60}C_1$	=	190	×	60	=	11,400	7.2
MATCH 3	$_{20}C_3$	×	$_{60}C_0$	=	1,140	×	1	=	1,140	72.1
					Possible outcomes $_{60}C_3$			=	82,160	

20/80 — Four numbers selected

	Favorable Outcomes	Probabilities
		1 in
Match 0:	487,635	3.2
Match 1:	684,400	2.3
Match 2:	336,300	4.7
Match 3:	68,400	23.1
Match 4:	4,845	326

	Winning numbers		Other numbers						Favorable outcomes	Probability of winning (1 in)
MATCH 0	$_{20}C_0$	×	$_{60}C_4$	=	1	×	487,635	=	487,635	3.2
MATCH 1	$_{20}C_1$	×	$_{60}C_3$	=	20	×	34,220	=	684,400	2.3
MATCH 2	$_{20}C_2$	×	$_{60}C_2$	=	190	×	1,770	=	336,300	4.7
MATCH 3	$_{20}C_3$	×	$_{60}C_1$	=	1,140	×	60	=	68,400	23.1
MATCH 4	$_{20}C_4$	×	$_{60}C_0$	=	4,845	×	1	=	4,845	326.4
						Possible outcomes $_{80}C_4$		=	1,581,580	

20/80 — Five numbers selected

	Favorable Outcomes	Probabilities (1 in)
Match 0:	5,461,512	4
Match 1:	9,752,700	2
Match 2:	6,501,800	4
Match 3:	2,017,800	12
Match 4:	290,700	83
Match 5:	15,504	1,551

	Winning numbers		Other numbers						Favorable outcomes	Probability of winning (1 in)
MATCH 0	$_{20}C_0$	×	$_{60}C_5$	=	1	×	5,461,512	=	5,461,512	4.4
MATCH 1	$_{20}C_1$	×	$_{60}C_4$	=	20	×	487,635	=	9,752,700	2.5
MATCH 2	$_{20}C_2$	×	$_{60}C_3$	=	190	×	34,220	=	6,501,800	3.7
MATCH 3	$_{20}C_3$	×	$_{60}C_2$	=	1,140	×	1,770	=	2,017,800	11.9
MATCH 4	$_{20}C_4$	×	$_{60}C_1$	=	4,845	×	60	=	290,700	82.7
MATCH 5	$_{20}C_5$	×	$_{60}C_0$	=	15,504	×	1	=	15,504	1,551
						Possible outcomes $_{80}C_5$		=	24,040,016	

20/80 — Six numbers selected

	Favorable Outcomes	Probabilities
		1 in
Match 0:	50,063,860	6
Match 1:	109,230,240	2.8
Match 2:	92,650,650	3.2
Match 3:	39,010,800	7.7
Match 4:	8,575,650	35
Match 5:	930,240	323
Match 6:	38,760	7,753

	Winning numbers		Other numbers						Favorable outcomes	Probability of winning
										1 in
MATCH 0	$_{20}C_0$	×	$_{60}C_6$	=	1	×	50,063,860	=	50,063,860	6.0
MATCH 1	$_{20}C_1$	×	$_{60}C_5$	=	20	×	5,461,512	=	109,230,240	2.8
MATCH 2	$_{20}C_2$	×	$_{60}C_4$	=	190	×	487,635	=	92,650,650	3.2
MATCH 3	$_{20}C_3$	×	$_{60}C_3$	=	1,140	×	34,220	=	39,010,800	7.7
MATCH 4	$_{20}C_4$	×	$_{60}C_2$	=	4,845	×	1,770	=	8,575,650	35.0
MATCH 5	$_{20}C_5$	×	$_{60}C_1$	=	15,504	×	60	=	930,240	323.0
MATCH 6	$_{20}C_6$	×	$_{60}C_0$	=	38,760	×	1	=	38,760	7,753

Possible outcomes $_{80}C_6$ = 300,500,200

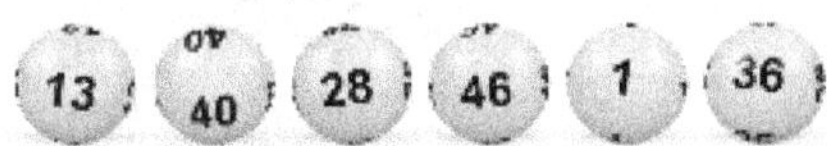

20/80 — Seven numbers selected

	Favorable Outcomes	Probabilities
		1 in
Match 0:	386,206,920	8.2
Match 1:	1,001,277,200	3.3
Match 2:	1,037,687,280	3.1
Match 3:	555,903,900	5.7
Match 4:	165,795,900	19
Match 5:	27,442,080	116
Match 6:	2,325,600	1,366
Match 7:	77,520	40,979

	Winning numbers		Other numbers						Favorable outcomes		Probability of winning
											1 in
MATCH 0	$_{20}C_0$	×	$_{60}C_7$	=	1	×	386,206,920	=	386,206,920		8.2
MATCH 1	$_{20}C_1$	×	$_{60}C_6$	=	20	×	50,063,860	=	1,001,277,200		3.2
MATCH 2	$_{20}C_2$	×	$_{60}C_5$	=	190	×	5,461,512	=	1,037,687,280		3.1
MATCH 3	$_{20}C_3$	×	$_{60}C_4$	=	1,140	×	487,635	=	555,903,900		5.7
MATCH 4	$_{20}C_4$	×	$_{60}C_3$	=	4,845	×	34,220	=	165,795,900		19.2
MATCH 5	$_{20}C_5$	×	$_{60}C_2$	=	15,504	×	1,770	=	27,442,080		115.8
MATCH 6	$_{20}C_6$	×	$_{60}C_1$	=	38,760	×	60	=	2,325,600		1,366
MATCH 7	$_{20}C_7$	×	$_{60}C_0$	=	77,520	×	1	=	77,520		40,979

				Possible outcomes		
				$_{80}C_7$	=	3,176,716,400

20/80 — Eight numbers selected

	Favorable Outcomes	Probabilities
		1 in
Match 0:	2,558,620,845	11.3
Match 1:	7,724,138,400	3.8
Match 2:	9,512,133,400	3.1
Match 3:	6,226,123,680	4.7
Match 4:	2,362,591,575	12.3
Match 5:	530,546,880	55
Match 6:	68,605,200	423
Match 7:	4,651,200	6,232
Match 8:	125,970	230,115

	Winning numbers		Other numbers						Favorable outcomes	Probability of winning
										1 in
MATCH 0	$_{20}C_0$	×	$_{60}C_8$	=	1	×	2,558,620,845	=	2,558,620,845	11.3
MATCH 1	$_{20}C_1$	×	$_{60}C_7$	=	20	×	386,206,920	=	7,724,138,400	3.8
MATCH 2	$_{20}C_2$	×	$_{60}C_6$	=	190	×	50,063,860	=	9,512,133,400	3.1
MATCH 3	$_{20}C_3$	×	$_{60}C_5$	=	1,140	×	5,461,512	=	6,226,123,680	4.7
MATCH 4	$_{20}C_4$	×	$_{60}C_4$	=	4,845	×	487,635	=	2,362,591,575	12.3
MATCH 5	$_{20}C_5$	×	$_{60}C_3$	=	15,504	×	34,220	=	530,546,880	54.6
MATCH 6	$_{20}C_6$	×	$_{60}C_2$	=	38,760	×	1,770	=	68,605,200	422.5
MATCH 7	$_{20}C_7$	×	$_{60}C_1$	=	77,520	×	60	=	4,651,200	6,232
MATCH 8	$_{20}C_8$	×	$_{60}C_0$	=	125,970	×	1	=	125,970	230,115
					Possible outcomes $_{80}C_8$			=	28,987,537,150	

20/80 — Nine numbers selected

	Favorable Outcomes	Probabilities
		1 in
Match 0:	14,783,142,660	15.7
Match 1:	51,172,416,900	4.5
Match 2:	73,379,314,800	3.2
Match 3:	57,072,800,400	4.1
Match 4:	26,461,025,640	8.8
Match 5:	7,560,293,040	31
Match 6:	1,326,367,200	175
Match 7:	137,210,400	1,690
Match 8:	7,558,200	30,682
Match 9:	167,960	1,380,688

	Winning numbers		Other numbers						Favorable outcomes		Probability of winning
											1 in
MATCH 0	$_{20}C_0$	×	$_{60}C_9$	=	1	×	14,783,142,660	=	14,783,142,660		15.7
MATCH 1	$_{20}C_1$	×	$_{60}C_8$	=	20	×	2,558,620,845	=	51,172,416,900		4.5
MATCH 2	$_{20}C_2$	×	$_{60}C_7$	=	190	×	386,206,920	=	73,379,314,800		3.2
MATCH 3	$_{20}C_3$	×	$_{60}C_6$	=	1,140	×	50,063,860	=	57,072,800,400		4.1
MATCH 4	$_{20}C_4$	×	$_{80}C_5$	=	4,845	×	5,461,512	=	26,461,025,640		8.8
MATCH 5	$_{20}C_5$	×	$_{60}C_4$	=	15,504	×	487,635	=	7,560,293,040		30.7
MATCH 6	$_{20}C_6$	×	$_{80}C_3$	=	38,760	×	34,220	=	1,326,367,200		174.8
MATCH 7	$_{20}C_7$	×	$_{60}C_2$	=	77,520	×	1,770	=	137,210,400		1,690
MATCH 8	$_{20}C_8$	×	$_{60}C_1$	=	125,970	×	60	=	7,558,200		30,682
MATCH 9	$_{20}C_9$	×	$_{60}C_0$	=	167,960	×	1	=	167,960		1,380,688

Possible outcomes $_{80}C_9$ = 231,900,297,200

20/80 — Ten numbers selected

	Favorable Outcomes	Probabilities
		1 in
Match 0:	75,394,027,566	21.8
Match 1:	295,662,853,200	5.6
Match 2:	486,137,960,550	3.4
Match 3:	440,275,888,800	3.7
Match 4:	242,559,401,700	6.8
Match 5:	84,675,282,048	19
Match 6:	18,900,732,600	87
Match 7:	2,652,734,400	621
Match 8:	222,966,900	7,384
Match 9:	10,077,600	163,381
Match 10:	184,756	8,911,711

	Winning numbers		Other numbers						Favorable outcomes	Probability of winning
										1 in
MATCH 0	$_{20}C_0$	×	$_{60}C_{10}$	=	1	×	75,394,027,566	=	75,394,027,566	21.8
MATCH 1	$_{20}C_1$	×	$_{60}C_9$	=	20	×	14,783,142,660	=	295,662,853,200	5.6
MATCH 2	$_{20}C_2$	×	$_{60}C_8$	=	190	×	2,558,620,845	=	486,137,960,550	3.4
MATCH 3	$_{20}C_3$	×	$_{60}C_7$	=	1,140	×	386,206,920	=	440,275,888,800	3.7
MATCH 4	$_{20}C_4$	×	$_{60}C_6$	=	4,845	×	50,063,860	=	242,559,401,700	6.8
MATCH 5	$_{20}C_5$	×	$_{60}C_5$	=	15,504	×	5,461,512	=	84,675,282,048	19.4
MATCH 6	$_{20}C_6$	×	$_{60}C_4$	=	38,760	×	487,635	=	18,900,732,600	87.1
MATCH 7	$_{20}C_7$	×	$_{60}C_3$	=	77,520	×	34,220	=	2,652,734,400	620.7
MATCH 8	$_{20}C_8$	×	$_{60}C_2$	=	125,970	×	1,770	=	222,966,900	7,385
MATCH 9	$_{20}C_9$	×	$_{60}C_1$	=	167,960	×	60	=	10,077,600	163,381
MATCH 10	$_{20}C_{10}$	×	$_{60}C_0$	=	184,756	×	1	=	184,756	8,911,711

Possible outcomes
$_{80}C_{10}$ = 1,646,492,110,120

20/80 — Eleven numbers selected

	Favorable Outcomes	Probabilities
		1 in
Match 0:	342,700,125,300	30.6
Match 1:	1,507,880,551,320	7
Match 2:	2,808,797,105,400	3.7
Match 3:	2,916,827,763,300	3.6
Match 4:	1,871,172,527,400	5.6
Match 5:	776,190,085,440	13.5
Match 6:	211,688,205,120	49.5
Match 7:	37,801,465,200	277
Match 8:	4,310,693,400	2,431
Match 9:	297,289,200	35,244
Match 10:	11,085,360	945,181
Match 11:	167,960	62,381,978

	Winning numbers		Other numbers						Favorable outcomes	Probability of winning
										1 in
MATCH 0	$_{20}C_0$	×	$_{60}C_{11}$	=	1	×	342,700,125,300	=	342,700,125,300	30.6
MATCH 1	$_{20}C_1$	×	$_{60}C_{10}$	=	20	×	75,394,027,566	=	1,507,880,551,320	7.0
MATCH 2	$_{20}C_2$	×	$_{80}C_9$	=	190	×	14,783,142,660	=	2,808,797,105,400	3.7
MATCH 3	$_{20}C_3$	×	$_{60}C_8$	=	1,140	×	2,558,620,845	=	2,916,827,763,300	3.6
MATCH 4	$_{20}C_4$	×	$_{60}C_7$	=	4,845	×	386,206,920	=	1,871,172,527,400	5.6
MATCH 5	$_{20}C_5$	×	$_{80}C_6$	=	15,504	×	50,063,860	=	776,190,085,440	13.5
MATCH 6	$_{20}C_6$	×	$_{80}C_5$	=	38,760	×	5,461,512	=	211,688,205,120	49.5
MATCH 7	$_{20}C_7$	×	$_{60}C_4$	=	77,520	×	487,635	=	37,801,465,200	277.2
MATCH 8	$_{20}C_8$	×	$_{60}C_3$	=	125,970	×	34,220	=	4,310,693,400	2,431
MATCH 9	$_{20}C_9$	×	$_{80}C_2$	=	167,960	×	1,770	=	297,289,200	35,244
MATCH 10	$_{20}C_{10}$	×	$_{80}C_1$	=	184,756	×	60	=	11,085,360	945,182
MATCH 11	$_{20}C_{11}$	×	$_{60}C_0$	=	167,960	×	1	=	167,960	62,381,978

Possible outcomes
$_{80}C_{11}$ = 10,477,677,064,400

20/80 — Twelve numbers selected

	Favorable Outcomes	Probabilities
		1 in
Match 0:	1,399,358,844,975	43.1
Match 1:	6,854,002,506,000	8.8
Match 2:	14,324,865,237,540	4.2
Match 3:	16,852,782,632,400	3.6
Match 4:	12,396,517,994,025	4.9
Match 5:	5,987,752,087,680	10
Match 6:	1,940,475,213,600	31
Match 7:	423,376,410,240	142
Match 8:	61,427,380,950	981
Match 9:	5,747,591,200	10,482
Match 10:	327,018,120	184,230
Match 11:	10,077,600	5,978,273
Match 12:	125,970	478,261,833

	Winning numbers		Other numbers					Favorable outcomes	Probability of winning
									1 in
MATCH 0	$_{20}C_0$	×	$_{60}C_{12}$	=	1	×	1,399,358,844,975 =	1,399,358,844,975	43.1
MATCH 1	$_{20}C_1$	×	$_{60}C_{11}$	=	20	×	342,700,125,300 =	6,854,002,506,000	8.8
MATCH 2	$_{20}C_2$	×	$_{60}C_{10}$	=	190	×	75,394,027,566 =	14,324,865,237,540	4.2
MATCH 3	$_{20}C_3$	×	$_{60}C_9$	=	1,140	×	14,783,142,660 =	16,852,782,632,400	3.6
MATCH 4	$_{20}C_4$	×	$_{60}C_8$	=	4,845	×	2,558,620,845 =	12,396,517,994,025	4.9
MATCH 5	$_{20}C_5$	×	$_{60}C_7$	=	15,504	×	386,206,920 =	5,987,752,087,680	10.1
MATCH 6	$_{20}C_6$	×	$_{60}C_6$	=	38,760	×	50,063,860 =	1,940,475,213,600	31.1
MATCH 7	$_{20}C_7$	×	$_{60}C_5$	=	77,520	×	5,461,512 =	423,376,410,240	142.3
MATCH 8	$_{20}C_8$	×	$_{60}C_4$	=	125,970	×	487,635 =	61,427,380,950	980.8
MATCH 9	$_{20}C_9$	×	$_{60}C_3$	=	167,960	×	34,220 =	5,747,591,200	10,482
MATCH 10	$_{20}C_{10}$	×	$_{60}C_2$	=	184,756	×	1,770 =	327,018,120	184,230
MATCH 11	$_{20}C_{11}$	×	$_{60}C_1$	=	167,960	×	60 =	10,077,600	5,978,273
MATCH 12	$_{20}C_{12}$	×	$_{60}C_0$	=	125,970	×	1 =	125,970	478,261,833

Possible outcomes
$_{60}C_{12}$ = 60,246,643,120,300

20/80 — Thirteen numbers selected

	Favorable Outcomes	Probabilities
		1 in
Match 0:	5,166,863,427,600	61
Match 1:	27,987,176,899,500	11
Match 2:	65,113,023,807,000	5
Match 3:	85,949,191,425,240	4
Match 4:	71,624,326,187,700	4
Match 5:	39,668,857,580,880	8
Match 6:	14,969,380,219,200	21
Match 7:	3,880,950,427,200	81
Match 8:	687,986,666,640	458
Match 9:	81,903,174,600	3,848
Match 10:	6,322,350,320	49,845
Match 11:	297,289,200	1,060,033
Match 12:	7,558,200	41,694,621
Match 13:	77,520	4,065,225,582

	Winning numbers		Other numbers						Favorable outcomes	Probability of winning
										1 in
MATCH 0	$_{20}C_0$	×	$_{60}C_{13}$	=	1	×	5,166,863,427,600	=	5,166,863,427,600	61.0
MATCH 1	$_{20}C_1$	×	$_{60}C_{12}$	=	20	×	1,399,358,844,975	=	27,987,176,899,500	11.3
MATCH 2	$_{20}C_2$	×	$_{60}C_{11}$	=	190	×	342,700,125,300	=	65,113,023,807,000	4.8
MATCH 3	$_{20}C_3$	×	$_{60}C_{10}$	=	1,140	×	75,394,027,566	=	85,949,191,425,240	3.7
MATCH 4	$_{20}C_4$	×	$_{60}C_9$	=	4,845	×	14,783,142,660	=	71,624,326,187,700	4.4
MATCH 5	$_{20}C_5$	×	$_{60}C_8$	=	15,504	×	2,558,620,845	=	39,668,857,580,880	7.9
MATCH 6	$_{20}C_6$	×	$_{60}C_7$	=	38,760	×	386,206,920	=	14,969,380,219,200	21.1
MATCH 7	$_{20}C_7$	×	$_{60}C_6$	=	77,520	×	50,063,860	=	3,880,950,427,200	81.2
MATCH 8	$_{20}C_8$	×	$_{60}C_5$	=	125,970	×	5,461,512	=	687,986,666,640	458.1
MATCH 9	$_{20}C_9$	×	$_{60}C_4$	=	167,960	×	487,635	=	81,903,174,600	3,848
MATCH 10	$_{20}C_{10}$	×	$_{60}C_3$	=	184,756	×	34,220	=	6,322,350,320	49,845
MATCH 11	$_{20}C_{11}$	×	$_{60}C_2$	=	167,960	×	1,770	=	297,289,200	1,060,033
MATCH 12	$_{20}C_{12}$	×	$_{60}C_1$	=	125,970	×	60	=	7,558,200	41,694,621
MATCH 13	$_{20}C_{13}$	×	$_{60}C_0$	=	77,520	×	1	=	77,520	4,065,225,582

		Possible outcomes	
	$_{80}C_{13}$	=	315,136,287,090,800

20/80 — Fourteen numbers selected

	Favorable Outcomes	Probabilities
		1 in
Match 0:	17,345,898,649,800	87
Match 1:	103,337,268,552,000	15
Match 2:	265,878,180,545,250	6
Match 3:	390,678,142,842,000	4
Match 4:	365,284,063,557,270	4
Match 5:	229,197,843,800,640	7
Match 6:	99,172,143,952,200	15
Match 7:	29,938,760,438,400	50
Match 8:	6,306,544,444,200	239
Match 9:	917,315,555,520	1,644
Match 10:	90,093,492,060	16,740
Match 11:	5,747,591,200	262,397
Match 12:	222,966,900	6,764,018
Match 13:	4,651,200	324,250,136
Match 14:	38,760	38,910,016,282

	Winning numbers		Other numbers						Favorable outcomes	Probability of winning
										1 in
MATCH 0	$_{20}C_0$	×	$_{60}C_{14}$	=	1	×	17,345,898,649,800	=	17,345,898,649,800	87.0
MATCH 1	$_{20}C_1$	×	$_{60}C_{13}$	=	20	×	5,166,863,427,600	=	103,337,268,552,000	14.6
MATCH 2	$_{20}C_2$	×	$_{60}C_{12}$	=	190	×	1,399,358,844,975	=	265,878,180,545,250	5.7
MATCH 3	$_{20}C_3$	×	$_{60}C_{11}$	=	1,140	×	342,700,125,300	=	390,678,142,842,000	3.9
MATCH 4	$_{20}C_4$	×	$_{60}C_{10}$	=	4,845	×	75,394,027,566	=	365,284,063,557,270	4.1
MATCH 5	$_{20}C_5$	×	$_{60}C_9$	=	15,504	×	14,783,142,660	=	229,197,843,800,640	6.6
MATCH 6	$_{20}C_6$	×	$_{60}C_8$	=	38,760	×	2,558,620,845	=	99,172,143,952,200	15.2
MATCH 7	$_{20}C_7$	×	$_{60}C_7$	=	77,520	×	386,206,920	=	29,938,760,438,400	50.4
MATCH 8	$_{20}C_8$	×	$_{60}C_6$	=	125,970	×	50,063,860	=	6,306,544,444,200	239.1
MATCH 9	$_{20}C_9$	×	$_{60}C_5$	=	167,960	×	5,461,512	=	917,315,555,520	1,644
MATCH 10	$_{20}C_{10}$	×	$_{60}C_4$	=	184,756	×	487,635	=	90,093,492,060	16,740
MATCH 11	$_{20}C_{11}$	×	$_{60}C_3$	=	167,960	×	34,220	=	5,747,591,200	262,397
MATCH 12	$_{20}C_{12}$	×	$_{60}C_2$	=	125,970	×	1,770	=	222,966,900	6,764,019
MATCH 13	$_{20}C_{13}$	×	$_{60}C_1$	=	77,520	×	60	=	4,651,200	324,250,136
MATCH 14	$_{20}C_{14}$	×	$_{60}C_0$	=	38,760	×	1	=	38,760	38,910,016,282

Possible outcomes

$$_{60}C_{14} = 1,508,152,231,077,400$$

20/80 — Fifteen numbers selected

	Favorable Outcomes	Probabilities
		1 in
Match 0:	53,194,089,192,720	125
Match 1:	346,917,972,996,000	19
Match 2:	981,704,051,244,000	7
Match 3:	1,595,269,083,271,500	4
Match 4:	1,660,382,107,078,500	4
Match 5:	1,168,909,003,383,260	6
Match 6:	572,994,609,501,600	12
Match 7:	198,344,287,904,400	33
Match 8:	48,650,485,712,400	136
Match 9:	8,408,725,925,600	789
Match 10:	1,009,047,111,072	6,576
Match 11:	81,903,174,600	81,021
Match 12:	4,310,693,400	1,539,397
Match 13:	137,210,400	48,362,732
Match 14:	2,325,600	2,853,401,194
Match 15:	15,504	428,010,179,098

	Winning numbers		Other numbers					Favorable outcomes	Probability of winning
									1 in
MATCH 0	$_{20}C_0$	×	$_{60}C_{15}$	=	1	×	53,194,089,192,720	= 53,194,089,192,720	124.8
MATCH 1	$_{20}C_1$	×	$_{60}C_{14}$	=	20	×	17,345,898,649,800	= 346,917,972,996,000	19.1
MATCH 2	$_{20}C_2$	×	$_{60}C_{13}$	=	190	×	5,166,863,427,600	= 981,704,051,244,000	6.8
MATCH 3	$_{20}C_3$	×	$_{80}C_{12}$	=	1,140	×	1,399,358,844,975	= 1,595,269,083,271,500	4.2
MATCH 4	$_{20}C_4$	×	$_{60}C_{11}$	=	4,845	×	342,700,125,300	= 1,660,382,107,078,500	4.0
MATCH 5	$_{20}C_5$	×	$_{60}C_{10}$	=	15,504	×	75,394,027,566	= 1,168,909,003,383,260	5.7
MATCH 6	$_{20}C_6$	×	$_{60}C_9$	=	38,760	×	14,783,142,660	= 572,994,609,501,600	11.6
MATCH 7	$_{20}C_7$	×	$_{60}C_8$	=	77,520	×	2,558,620,845	= 198,344,287,904,400	33.5
MATCH 8	$_{20}C_8$	×	$_{60}C_7$	=	125,970	×	386,206,920	= 48,650,485,712,400	136.4
MATCH 9	$_{20}C_9$	×	$_{60}C_6$	=	167,960	×	50,063,860	= 8,408,725,925,600	789.2
MATCH 10	$_{20}C_{10}$	×	$_{60}C_5$	=	184,756	×	5,461,512	= 1,009,047,111,072	6,576
MATCH 11	$_{20}C_{11}$	×	$_{60}C_4$	=	167,960	×	487,635	= 81,903,174,600	81,021
MATCH 12	$_{20}C_{12}$	×	$_{60}C_3$	=	125,970	×	34,220	= 4,310,693,400	1,539,397
MATCH 13	$_{20}C_{13}$	×	$_{60}C_2$	=	77,520	×	1,770	= 137,210,400	48,362,732
MATCH 14	$_{20}C_{14}$	×	$_{60}C_1$	=	38,760	×	60	= 2,325,600	2,853,401,194
MATCH 15	$_{20}C_{15}$	×	$_{60}C_0$	=	15,504	×	1	= 15,504	428,010,179,008

Possible outcomes
$_{80}C_{15}$ = 6,635,869,816,740,560

Probabilities of all Lotto 20/80 divisions

(The column "Prize" indicates if a prize is paid in Australia)

Player's selection	Division	Probability 1 in	Prize
Select 1	Match 0	1.3	No
Select 2	Match 0	1.8	No
Select 4	Match 1	2.3	No
Select 3	Match 1	2.3	No
Select 3	Match 0	2.4	No
Select 5	Match 1	2.5	No
Select 2	Match 1	2.6	No
Select 6	Match 1	2.8	No
Select 8	Match 2	3.1	No
Select 7	Match 2	3.1	No
Select 9	Match 2	3.2	No
Select 7	Match 1	3.2	No
Select 4	Match 0	3.2	No
Select 6	Match 2	3.2	No
Select 10	Match 2	3.4	No
Select 12	Match 3	3.6	No
Select 11	Match 3	3.6	No
Select 13	Match 3	3.7	No
Select 5	Match 2	3.7	No
Select 11	Match 2	3.7	No
Select 10	Match 3	3.7	No
Select 8	Match 1	3.8	No
Select 14	Match 3	3.9	No
Select 15	Match 4	4.0	No
Select 1	Match 1	4.0	Yes
Select 9	Match 3	4.1	No

Select 14	Match 4	4.1	No
Select 15	Match 3	4.2	No
Select 12	Match 2	4.2	No
Select 20	Match 5	4.3	No
Select 13	Match 4	4.4	No
Select 5	Match 0	4.4	No
Select 9	Match 1	4.5	No
Select 8	Match 3	4.7	No
Select 4	Match 2	4.7	Yes
Select 13	Match 2	4.8	No
Select 12	Match 4	4.9	No
Select 20	Match 4	4.9	No
Select 40	Match 10	4.9	No
Select 20	Match 6	5.3	No
Select 10	Match 1	5.6	No
Select 40	Match 11	5.6	No
Select 40	Match 9	5.6	No
Select 11	Match 4	5.6	No
Select 14	Match 2	5.7	No
Select 15	Match 5	5.7	Yes
Select 7	Match 3	5.7	No
Select 6	Match 0	6.0	No
Select 14	Match 5	6.6	No
Select 15	Match 2	6.8	No
Select 10	Match 4	6.8	Yes
Select 11	Match 1	7.0	No
Select 3	Match 2	7.2	Yes
Select 6	Match 3	7.7	Yes
Select 13	Match 5	7.9	No
Select 20	Match 3	8.0	No
Select 7	Match 0	8.2	No
Select 40	Match 12	8.2	No

Select 40	Match 8	8.2	No
Select 9	Match 4	8.8	Yes
Select 12	Match 1	8.8	No
Select 20	Match 7	8.8	No
Select 12	Match 5	10.1	No
Select 13	Match 1	11.3	No
Select 8	Match 0	11.3	No
Select 15	Match 6	11.6	Yes
Select 5	Match 3	11.9	Yes
Select 8	Match 4	12.3	Yes
Select 11	Match 5	13.5	No
Select 14	Match 1	14.6	No
Select 14	Match 6	15.2	No
Select 9	Match 0	15.7	No
Select 40	Match 13	15.8	Yes
Select 40	Match 7	15.8	Yes
Select 2	Match 2	16.6	Yes
Select 15	Match 1	19.1	No
Select 7	Match 4	19.2	Yes
Select 10	Match 5	19.4	Yes
Select 20	Match 8	20.1	Yes
Select 20	Match 2	20.1	Yes
Select 13	Match 6	21.1	No
Select 10	Match 0	21.8	No
Select 4	Match 3	23.1	Yes
Select 11	Match 0	30.6	No
Select 9	Match 5	30.7	Yes
Select 12	Match 6	31.1	No
Select 15	Match 7	33.5	Yes
Select 6	Match 4	35.0	Yes
Select 40	Match 14	39.7	Yes
Select 40	Match 6	39.7	Yes

Select 12	Match 0	43.1	No
Select 11	Match 6	49.5	No
Select 14	Match 7	50.4	No
Select 8	Match 5	54.6	Yes
Select 13	Match 0	61.0	No
Select 20	Match 9	61.4	Yes
Select 3	Match 3	72.1	Yes
Select 13	Match 7	81.2	No
Select 5	Match 4	82.7	Yes
Select 20	Match 1	86.4	Yes
Select 14	Match 0	87.0	No
Select 10	Match 6	87.1	Yes
Select 7	Match 5	115.8	Yes
Select 15	Match 0	124.8	Yes
Select 40	Match 15	133.6	Yes
Select 40	Match 5	133.6	Yes
Select 15	Match 8	136.4	Yes
Select 12	Match 7	142.3	Yes
Select 9	Match 6	174.8	Yes
Select 14	Match 8	239.1	Yes
Select 20	Match 10	253.8	Yes
Select 11	Match 7	277.2	Yes
Select 6	Match 5	323.0	Yes
Select 4	Match 4	326.4	Yes
Select 8	Match 6	422.5	Yes
Select 13	Match 8	458.1	Yes
Select 40	Match 16	615.5	Yes
Select 40	Match 4	615.5	Yes
Select 10	Match 7	620.7	Yes
Select 15	Match 9	789.2	Yes
Select 20	Match 0	843.4	Yes
Select 12	Match 8	980.8	Yes

Select 7	Match 6	1,366	Yes
Select 20	Match 11	1,424	Yes
Select 5	Match 5	1,551	Yes
Select 14	Match 9	1,644	Yes
Select 9	Match 7	1,690	Yes
Select 11	Match 8	2,431	Yes
Select 13	Match 9	3,848	Yes
Select 40	Match 17	4,033	Yes
Select 40	Match 3	4,033	Yes
Select 8	Match 7	6,232	Yes
Select 15	Match 10	6,576	Yes
Select 10	Match 8	7,385	Yes
Select 6	Match 6	7,753	Yes
Select 12	Match 9	10,482	Yes
Select 20	Match 12	10,969	Yes
Select 14	Match 10	16,740	Yes
Select 9	Match 8	30,682	Yes
Select 11	Match 9	35,244	Yes
Select 40	Match 18	39,976	Yes
Select 40	Match 2	39,976	Yes
Select 7	Match 7	40,979	Yes
Select 13	Match 10	49,845	Yes
Select 15	Match 11	81,021	Yes
Select 20	Match 13	118,085	Yes
Select 10	Match 9	163,381	Yes
Select 12	Match 10	184,230	Yes
Select 8	Match 8	230,115	Yes
Select 14	Match 11	262,397	Yes
Select 40	Match 19	673,227	Yes
Select 40	Match 1	673,227	Yes
Select 11	Match 10	945,182	Yes
Select 13	Match 11	1,060,033	Yes

Select 9	Match 9	1,380,688	Yes
Select 15	Match 12	1,539,397	Yes
Select 20	Match 14	1,821,882	Yes
Select 12	Match 11	5,978,273	Yes
Select 14	Match 12	6,764,019	Yes
Select 10	Match 10	8,911,711	Yes
Select 40	Match 20	25,646,755	Yes
Select 40	Match 0	25,646,755	Yes
Select 13	Match 12	41,694,621	Yes
Select 20	Match 15	41,751,454	Yes
Select 15	Match 13	48,362,732	Yes
Select 11	Match 11	62,381,978	Yes
Select 14	Match 13	324,250,136	Yes
Select 12	Match 12	478,261,833	Yes
Select 20	Match 16	1,496,372,111	Yes
Select 15	Match 14	2,853,401,194	Yes
Select 13	Match 13	4,065,225,582	Yes
Select 14	Match 14	38,910,016,282	Yes
Select 20	Match 17	90,624,035,965	Yes
Select 15	Match 15	428,010,179,098	Yes
Select 20	Match 18	10,512,388,171,907	Yes
Select 20	Match 19	2,946,096,785,176,810	Yes
Select 20	Match 20	3,535,316,142,212,170,000	Yes

CHAPTER 6

DRAWS WITHOUT REPLACEMENT
TWO DRAW MACHINES

Draws without replacement include lotteries in which every number drawn is not returned into the system before drawing the next number. As it could be expected this procedure does not produce repeated numbers. Because order is not important in this case, most of the following problems are solved by using the formulas applied to combinations.

In these draws, the player selects five, six, or seven numbers from one to a certain maximum—specified by the lottery. They are usually denoted by the abbreviation n/N, which means that you must select n numbers from 1 to N. A typical example is 6/45, where you select six numbers from 1 to 45.

In this chapter we are going to study lotteries that use two machines. Those lotteries have the structure m/M plus n/N. Winning the Division 1 prize in this type of lotteries is equivalent to winning two lotteries simultaneously because you must match the numbers drawn from the first machine, as well as those drawn from the second machine.

The formulas applied to calculate the probabilities on this type of draws are those explained in Chapter 2, under the subtitle *Compound Events*.

NOTE:

Some lotteries offer an additional number known as a bonus, or supplementary number. In this book, we will use the abbreviation "suppl." to refer to this extra number.

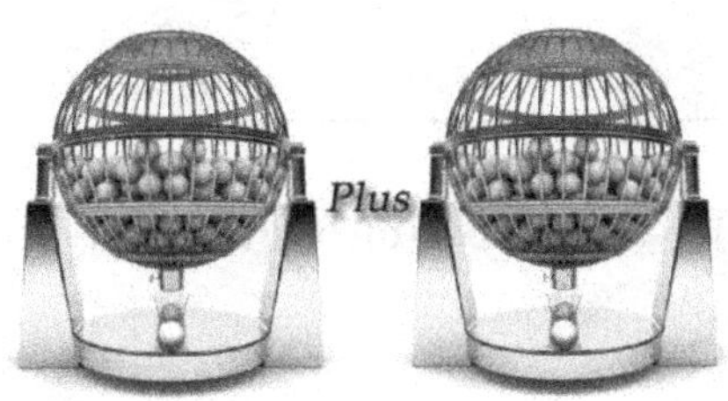

2/26 Plus 2/26

The first lottery drawing machine (Machine A) contains twenty-six balls numbered from 1 to 26. Two balls are drawn without replacement. These are the machine A winning numbers.

The second lottery drawing machine (Machine B) contains twenty-six balls numbered from 1 to 26. Two balls are drawn without replacement. These are the machine B winning numbers.

MACHINE A	MACHINE B
A B	P Q
2/26	2/26

Prize Structure and Probabilities

2/26 plus 2/26

	Match Machine A Winning Numbers	Machine B Winning Numbers	Favorable Outcomes	Probabilities
				1 in
DIV 1	2	2	1	105,625
DIV 2	2	1	48	2,201
DIV 3	1	2	48	2,201
DIV 4	2		276	383
DIV 5		2	276	383
DIV 6	1	1	2,304	46
DIV 7	1		13,248	8
DIV 8		1	13,248	8

Total Possible Outcomes

$$C(26,2) \times C(26,2) = 325 \times 325 = 105,625$$

Calculation of Favorable Outcomes

Match A	Match B	MACHINE A W		MACHINE B W		W = Winning numbers
2	2	$C_2^2 \times C_0^{24}$	$\times$	$C_2^2 \times C_0^{24}$		$= 1 \times 1 \times 1 \times 1 = 1$
2	1	$C_2^2 \times C_0^{24}$	$\times$	$C_1^2 \times C_1^{24}$		$= 1 \times 1 \times 2 \times 24 = 48$
1	2	$C_1^2 \times C_1^{24}$	$\times$	$C_2^2 \times C_0^{24}$		$= 2 \times 24 \times 1 \times 1 = 48$
2	0	$C_2^2 \times C_0^{24}$	$\times$	$C_0^2 \times C_2^{24}$		$= 1 \times 1 \times 1 \times 276 = 276$
0	2	$C_0^2 \times C_2^{24}$	$\times$	$C_2^2 \times C_0^{24}$		$= 1 \times 276 \times 1 \times 1 = 276$
1	1	$C_1^2 \times C_1^{24}$	$\times$	$C_1^2 \times C_1^{24}$		$= 2 \times 24 \times 2 \times 24 = 2{,}304$
1	0	$C_1^2 \times C_1^{24}$	$\times$	$C_0^2 \times C_2^{24}$		$= 2 \times 24 \times 1 \times 276 = 13{,}248$
0	1	$C_0^2 \times C_2^{24}$	$\times$	$C_1^2 \times C_1^{24}$		$= 1 \times 276 \times 2 \times 24 = 13{,}248$

A lottery of this type is played in Kansas,
Nebraska, North Dakota, and Wyoming, USA.

4/35 Plus 1/25

The lottery drawing machine (Machine A) contains thirty-five balls numbered from 1 to 35. Four balls are drawn without replacement. These are the machine A winning numbers.

A second lottery drawing machine (Machine B) contains twenty-five balls numbered from 1 to 25. One ball is drawn without replacement. This is the machine B winning number.

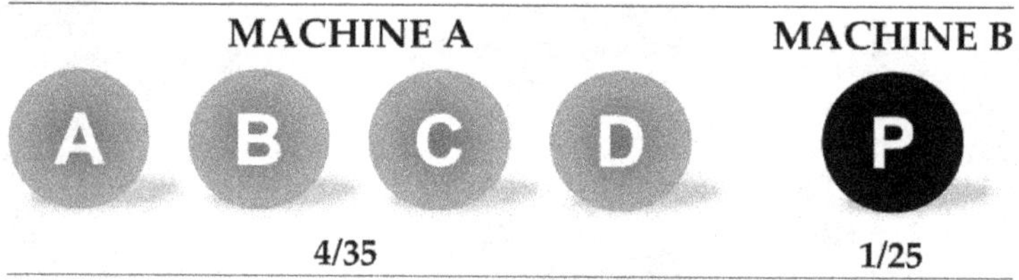

Prize Structure and Probabilities

	4/35 plus 1/25			
		Match		
	Machine **A**	Machine **B**	Favorable Outcomes	Probabilities
	Winning Numbers	Winning Number		
				1 in
DIV 1	4	1	1	1,309,000
DIV 2	4		24	54,542
DIV 3	3	1	124	10,556
DIV 4	3		2,976	440
DIV 5	2	1	2,790	469
DIV 6	2		66,960	20
DIV 7	1	1	17,980	73
DIV 8		1	31,465	42

Total Possible Outcomes

$$C(35,4) \times C(25,1) = 52,360 \times 25 = 1,309,000$$

Calculation of Favorable Outcomes

Match A	Match B	MACHINE A		MACHINE B		W = Winning numbers	
4	1	C^4_4 x	C^{31}_0 x	C^1_1 x	C^{24}_0	= 1 x 1 x 1 x 1 =	1
4	0	C^4_4 x	C^{31}_0 x	C^1_0 x	C^{24}_1	= 1 x 1 x 1 x 24 =	24
3	1	C^4_3 x	C^{31}_1 x	C^1_1 x	C^{24}_0	= 4 x 31 x 1 x 1 =	124
3	0	C^4_3 x	C^{31}_1 x	C^1_0 x	C^{24}_1	= 4 x 31 x 1 x 24 =	2,976
2	1	C^4_2 x	C^{31}_2 x	C^1_1 x	C^{24}_0	= 6 x 465 x 1 x 1 =	2,790
2	0	C^4_2 x	C^{31}_2 x	C^1_0 x	C^{24}_1	= 6 x 465 x 1 x 24 =	66,960
1	1	C^4_1 x	C^{31}_3 x	C^1_1 x	C^{24}_0	= 4 x 4,495 x 1 x 1 =	17,980
0	1	C^4_0 x	C^{31}_4 x	C^1_1 x	C^{24}_0	= 1 x 31,465 x 1 x 1 =	31,465

A lottery of this type is played in Kentucky, USA.

4/35 Plus 1/35

The lottery drawing machine (Machine A) contains thirty-five balls numbered from 1 to 35. Four balls are drawn without replacement. These are the machine A winning numbers.

A second lottery drawing machine (Machine B) contains thirty-five balls numbered from 1 to 35. One ball is drawn without replacement. This is the machine B winning number.

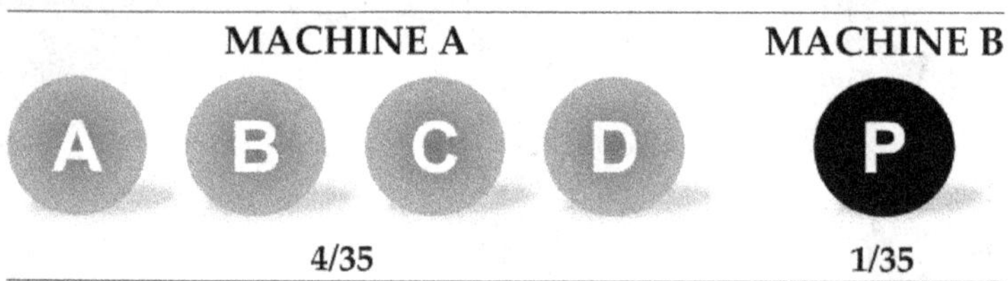

Prize Structure and Probabilities

4/35 plus 1/35

	Match			
	Machine **A** Winning Numbers	Machine **B** Winning Number	Favorable Outcomes	Probabilities
				1 in
DIV 1	4	1	1	1,832,600
DIV 2	4		34	53,900
DIV 3	3	1	124	14,779
DIV 4	3		4,216	435
DIV 5	2	1	2,790	657
DIV 6	1	1	17,980	102
DIV 7		1	31,465	58

Total Possible Outcomes

$$C(35,4) \times C(35,1) = 52{,}360 \times 35 = 1{,}832{,}600$$

Calculation of Favorable Outcomes

Match A	Match B	MACHINE A W		MACHINE B W		W = Winning numbers	
4	1	C_4^4 ×	C_0^{31} ×	C_1^1 ×	C_0^{34}	= 1 x 1 x 1 x 1 =	1
4	0	C_4^4 ×	C_0^{31} ×	C_0^1 ×	C_1^{34}	= 1 x 1 x 1 x 34 =	34
3	1	C_3^4 ×	C_1^{31} ×	C_1^1 ×	C_0^{34}	= 4 x 31 x 1 x 1 =	124
3	0	C_3^4 ×	C_1^{31} ×	C_0^1 ×	C_1^{34}	= 4 x 31 x 1 x 34 =	4,216
2	1	C_2^4 ×	C_2^{31} ×	C_1^1 ×	C_0^{34}	= 6 x 465 x 1 x 1 =	2,790
1	1	C_1^4 ×	C_3^{31} ×	C_1^1 ×	C_0^{34}	= 4 x 4,495 x 1 x 1 =	17,980
0	1	C_0^4 ×	C_4^{31} ×	C_1^1 ×	C_0^{34}	= 1 x 31,465 x 1 x 1 =	31,465

A lottery of this type is played in Texas, USA.

5/32 Plus 1/25

The lottery drawing machine (Machine A) contains thirty-two balls numbered from 1 to 32. Five balls are drawn without replacement. These are the machine A winning numbers.

A second lottery drawing machine (Machine B) contains twenty-five balls numbered from 1 to 25. One ball is drawn without replacement. This is the machine B winning number.

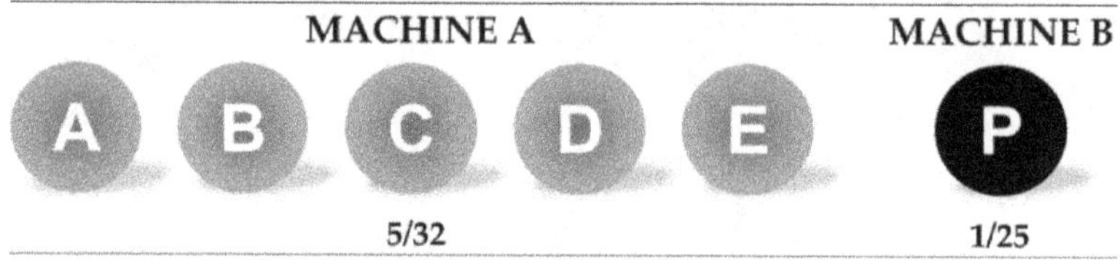

Prize Structure and Probabilities

5/32 plus 1/25
Prize Structure and Probabilities

	Match			Probabilities	
	Machine A Winning Numbers	Machine B Winning Number	Favorable Outcomes	1 game 1 in	2 games 1 in
DIV 1	5	1	1	5,034,400	2,517,200
DIV 2	5		24	209,767	104,884
DIV 3	4	1	135	37,292	18,646
DIV 4	4		3,240	1,554	777
DIV 5	3	1	3,510	1,434	717
DIV 6	3		84,240	60	30
DIV 7	2	1	29,250	172	86
DIV 8	1	1	87,750	57	29

Total Possible Outcomes

$$C(32,5) \times C(25,1) = 201,376 \times 25 = 5,034,400$$

Calculation of Favorable Outcomes

Match		MACHINE A		MACHINE B		W = Winning numbers
A	**B**	**W**		**W**		
5	1	C_5^5 x C_0^{27}	x	C_1^1 x C_0^{24}	$= 1 \times 1 \times 1 \times 1 =$	1
5	0	C_5^5 x C_0^{27}	x	C_0^1 x C_1^{24}	$= 1 \times 1 \times 1 \times 24 =$	24
4	1	C_4^5 x C_1^{27}	x	C_1^1 x C_0^{24}	$= 5 \times 27 \times 1 \times 1 =$	135
4	0	C_4^5 x C_1^{27}	x	C_0^1 x C_1^{24}	$= 5 \times 27 \times 1 \times 24 =$	3,240
3	1	C_3^5 x C_2^{27}	x	C_1^1 x C_0^{24}	$= 10 \times 351 \times 1 \times 1 =$	3,510
3	0	C_3^5 x C_2^{27}	x	C_0^1 x C_1^{24}	$= 10 \times 351 \times 1 \times 24 =$	84,240
2	1	C_2^5 x C_3^{27}	x	C_1^1 x C_0^{24}	$= 10 \times 2,925 \times 1 \times 1 =$	29,250
1	1	C_1^5 x C_4^{27}	x	C_1^1 x C_0^{24}	$= 5 \times 17,550 \times 1 \times 1 =$	87,750

A lottery of this type is played in Kansas, USA.

5/35 Plus 1/5

The lottery drawing machine (Machine A) contains thirty-five balls numbered from 1 to 35. Five balls are drawn without replacement. These are the machine A winning numbers.

A second lottery drawing machine (Machine B) contains five balls numbered from 1 to 5. One ball is drawn without replacement. This is the machine B winning number.

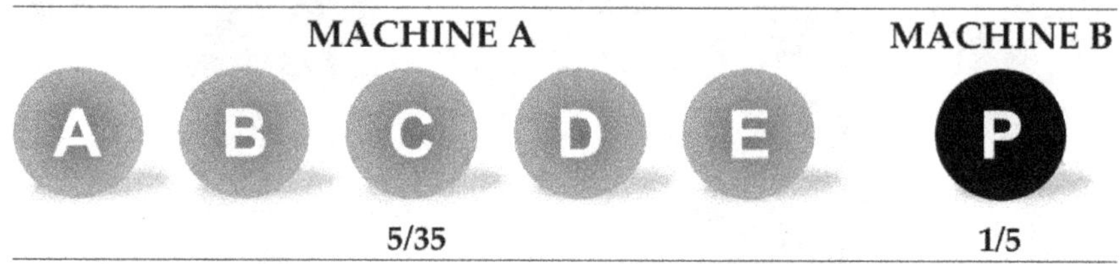

Prize Structure and Probabilities

5/35 plus 1/5

	Match			
	Machine **A** Winning Numbers	Machine **B** Winning Number	Favorable Outcomes	Probabilities
				1 in
DIV 1	5	1	1	1,623,160
DIV 2	5		4	405,790
DIV 3	4	1	150	10,821
DIV 4	4		600	2,705
DIV 5	3	1	4,350	373
DIV 6	3		17,400	93
DIV 7	2	1	40,600	40
DIV 8	2		162,400	10

Total Possible Outcomes

$$C(35,5) \times C(5,1) = 324,632 \times 5 = 1,623,160$$

Calculation of Favorable Outcomes

Match A	Match B	MACHINE A — W		MACHINE B — W		W = Winning numbers
5	1	C_5^5 ×	C_0^{30} ×	C_1^1 ×	C_0^4	= 1 x 1 x 1 x 1 = 1
5	0	C_5^5 ×	C_0^{30} ×	C_0^1 ×	C_1^4	= 1 x 1 x 1 x 4 = 4
4	1	C_4^5 ×	C_1^{30} ×	C_1^1 ×	C_0^4	= 5 x 30 x 1 x 1 = 150
4	0	C_4^5 ×	C_1^{30} ×	C_0^1 ×	C_1^4	= 5 x 30 x 1 x 4 = 600
3	1	C_3^5 ×	C_2^{30} ×	C_1^1 ×	C_0^4	= 10 x 435 x 1 x 1 = 4,350
3	0	C_3^5 ×	C_2^{30} ×	C_0^1 ×	C_1^4	= 10 x 435 x 1 x 4 = 17,400
2	1	C_2^5 ×	C_3^{30} ×	C_1^1 ×	C_0^4	= 10 x 4,060 x 1 x 1 = 40,600
2	0	C_2^5 ×	C_3^{30} ×	C_0^1 ×	C_1^4	= 10 x 4,060 x 1 x 4 = 162,400

A lottery of this type is played in Tennessee, USA.

5/39 Plus 1/14

The lottery drawing machine (Machine A) contains thirty-nine balls numbered from 1 to 39. Five balls are drawn without replacement. These are the machine A winning numbers.

A second lottery drawing machine (Machine B) contains fourteen balls numbered from 1 to 14. One ball is drawn without replacement. This is the machine B winning number.

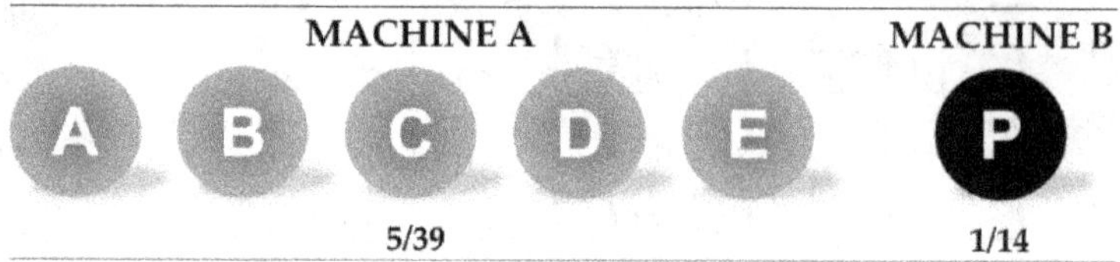

Prize Structure and Probabilities

5/39 plus 1/14

	Match Machine A Winning Numbers	Machine B Winning Number	Favorable Outcomes	Probabilities
				1 in
DIV 1	5	1	1	8,060,598
DIV 2	5		13	620,046
DIV 3	4	1	170	47,415
DIV 4	4		2,210	3,647
DIV 5	3	1	5,610	1,437
DIV 6	3		72,930	111
DIV 7	2	1	59,840	135
DIV 8	1	1	231,880	35
DIV 9		1	278,256	29

Total Possible Outcomes

$$C(39,5) \times C(14,1) = 575,757 \times 14 = 8,060,598$$

Calculation of Favorable Outcomes

Match A	B	MACHINE A: W		MACHINE B: W		W = Winning numbers	
5	1	C_5^5 ×	C_0^{34} ×	C_1^1 ×	C_0^{13}	= 1 x 1 x 1 x 1 =	1
5	0	C_5^5 ×	C_0^{34} ×	C_0^1 ×	C_1^{13}	= 1 x 1 x 1 x 13 =	13
4	1	C_4^5 ×	C_1^{34} ×	C_1^1 ×	C_0^{13}	= 5 x 34 x 1 x 1 =	170
4	0	C_4^5 ×	C_1^{34} ×	C_0^1 ×	C_1^{13}	= 5 x 34 x 1 x 13 =	2,210
3	1	C_3^5 ×	C_2^{34} ×	C_1^1 ×	C_0^{13}	= 10 x 561 x 1 x 1 =	5,610
3	0	C_3^5 ×	C_2^{34} ×	C_0^1 ×	C_1^{13}	= 10 x 561 x 1 x 13 =	72,930
2	1	C_2^5 ×	C_3^{34} ×	C_1^1 ×	C_0^{13}	= 10 x 5,984 x 1 x 1 =	59,840
1	1	C_1^5 ×	C_4^{34} ×	C_1^1 ×	C_0^{13}	= 5 x 46,376 x 1 x 1 =	231,880
0	1	C_0^5 ×	C_5^{34} ×	C_1^1 ×	C_0^{13}	= 1 x 278,256 x 1 x 1 =	278,256

A lottery of this type is played in the United Kingdom.

5/41 Plus 1/6

The lottery drawing machine (Machine A) contains forty-one balls numbered from 1 to 41. Five balls are drawn without replacement. Those are the machine A winning numbers.

A second lottery drawing machine (Machine B) contains six balls numbered from 1 to 6. One ball is drawn without replacement. This is the machine B winning number.

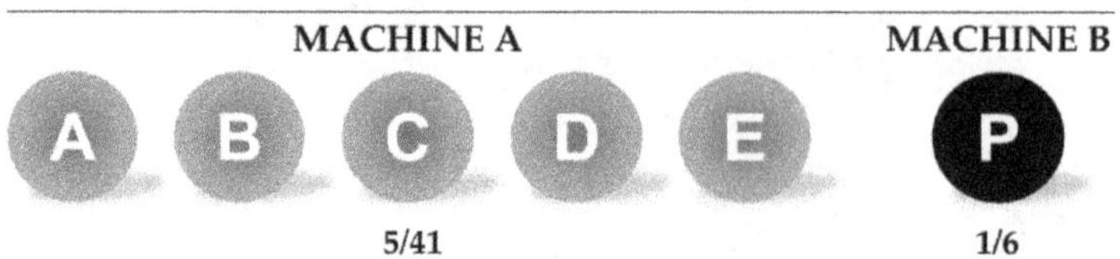

Prize Structure and Probabilities

5/41 plus 1/6

	Match			
	Machine **A** Winning Numbers	Machine **B** Winning Number	Favorable Outcomes	Probabilities
				1 in
DIV 1	5	1	1	4,496,388
DIV 2	5		5	899,278
DIV 3	4	1	180	24,980
DIV 4	4		900	4,996
DIV 5	3	1	6,300	714
DIV 6	3		31,500	143
DIV 7	2	1	71,400	63
DIV 8	2		357,000	13
DIV 9	1	1	294,525	15

Total Possible Outcomes

$$C(41,5) \times C(6,1) = 749,398 \times 6 = 4,496,388$$

Calculation of Favorable Outcomes

Match A	Match B	MACHINE A (W)		MACHINE B (W)		W = winning numbers	
5	1	C_5^5 x	C_0^{36} x	C_1^1 x	C_0^5	$= 1 \times 1 \times 1 \times 1 =$	1
5	0	C_5^5 x	C_0^{36} x	C_0^1 x	C_1^5	$= 1 \times 1 \times 1 \times 5 =$	5
4	1	C_4^5 x	C_1^{36} x	C_1^1 x	C_0^5	$= 5 \times 36 \times 1 \times 1 =$	180
4	0	C_4^5 x	C_1^{36} x	C_0^1 x	C_1^5	$= 5 \times 36 \times 1 \times 5 =$	900
3	1	C_3^5 x	C_2^{36} x	C_1^1 x	C_0^5	$= 10 \times 630 \times 1 \times 1 =$	6,300
3	0	C_3^5 x	C_2^{36} x	C_0^1 x	C_1^5	$= 10 \times 630 \times 1 \times 5 =$	31,500
2	1	C_2^5 x	C_3^{36} x	C_1^1 x	C_0^5	$= 10 \times 7,140 \times 1 \times 1 =$	71,400
2	0	C_2^5 x	C_3^{36} x	C_0^1 x	C_1^5	$= 10 \times 7,140 \times 1 \times 5 =$	357,000
1	1	C_1^5 x	C_4^{36} x	C_1^1 x	C_0^5	$= 5 \times 58,905 \times 1 \times 1 =$	294,525

A lottery of this type is played in Maine,
New Hampshire, and Vermont, USA.

5/43 Plus 1/16

The lottery drawing machine (Machine A) contains forty-three balls numbered from 1 to 43. Five balls are drawn without replacement. Those are the machine A winning numbers.

A second lottery drawing machine (Machine B) contains sixteen balls numbered from 1 to 16. One ball is drawn without replacement. This is the machine B winning number.

Prize Structure and Probabilities

5/43 plus 1/16

	Match			
	Machine A Winning Numbers	Machine B Winning Number	Favorable Outcomes	Probabilities
				1 in
DIV 1	5	1	1	15,401,568
DIV 2	5		15	1,026,771
DIV 3	4	1	190	81,061
DIV 4	4		2,850	5,404
DIV 5	3	1	7,030	2,191
DIV 6	3		105,450	146
DIV 7	2	1	84,360	183
DIV 8	1	1	369,075	42
DIV 9		1	501,942	31

Total Possible Outcomes

$$C(43,5) \times C(16,1) = 962{,}598 \times 16 = 15{,}401{,}568$$

Calculation of Favorable Outcomes

Match A	Match B	MACHINE A W		MACHINE B W		W = Winning numbers	
5	1	C_5^5 ×	C_0^{38} ×	C_1^1 ×	C_0^{15}	= 1 x 1 x 1 x 1 =	1
5	0	C_5^5 ×	C_0^{38} ×	C_0^1 ×	C_1^{15}	= 1 x 1 x 1 x 15 =	15
4	1	C_4^5 ×	C_1^{38} ×	C_1^1 ×	C_0^{15}	= 5 x 38 x 1 x 1 =	190
4	0	C_4^5 ×	C_1^{38} ×	C_0^1 ×	C_1^{15}	= 5 x 38 x 1 x 15 =	2,850
3	1	C_3^5 ×	C_2^{38} ×	C_1^1 ×	C_0^{15}	= 10 x 703 x 1 x 1 =	7,030
3	0	C_3^5 ×	C_2^{38} ×	C_0^1 ×	C_1^{15}	= 10 x 703 x 1 x 15 =	105,450
2	1	C_2^5 ×	C_3^{38} ×	C_1^1 ×	C_0^{15}	= 10 x 8,436 x 1 x 1 =	84,360
1	1	C_1^5 ×	C_4^{38} ×	C_1^1 ×	C_0^{15}	= 5 x 73,815 x 1 x 1 =	369,075
0	1	C_0^5 ×	C_5^{38} ×	C_1^1 ×	C_0^{15}	= 1 x 501,942 x 1 x 1 =	501,942

A lottery of this type is played in Colombia.

5/47 Plus 1/10

A lottery drawing machine (Machine A) contains forty-seven balls numbered from 1 to 47. Five balls are drawn without replacement. These are the machine A winning numbers.

A second lottery drawing machine (Machine B) contains ten balls numbered from 1 to 10. One ball is drawn without replacement. This is the machine B winning number.

Prize Structure and Probabilities

5/47 plus 1/10

	Match			
	Machine A Winning Numbers	Machine B Winning Number	Favorable Outcomes	Probabilities
				1 in
DIV 1	5	1	1	15,339,390
DIV 2	5		9	1,704,377
DIV 3	4	1	210	73,045
DIV 4	4		1,890	8,116
DIV 5	3	1	8,610	1,782
DIV 6	3		77,490	198
DIV 7	2	1	114,800	134
DIV 8	2		1,033,200	15

Total Possible Outcomes

$$C(47,5) \times C(10,1) = 1,533,939 \times 10 = 15,339,390$$

Calculation of Favorable Outcomes

Match A	B	MACHINE A W		MACHINE B W		W = Winning numbers	
5	1	C_5^5 $\times$	C_0^{42} $\times$	C_1^1 $\times$	C_0^9	$= 1 \times 1 \times 1 \times 1 =$	1
5	0	C_5^5 $\times$	C_0^{42} $\times$	C_0^1 $\times$	C_1^9	$= 1 \times 1 \times 1 \times 9 =$	9
4	1	C_4^5 $\times$	C_1^{42} $\times$	C_1^1 $\times$	C_0^9	$= 5 \times 42 \times 1 \times 1 =$	210
4	0	C_4^5 $\times$	C_1^{42} $\times$	C_0^1 $\times$	C_1^9	$= 5 \times 42 \times 1 \times 9 =$	1,890
3	1	C_3^5 $\times$	C_2^{42} $\times$	C_1^1 $\times$	C_0^9	$= 10 \times 861 \times 1 \times 1 =$	8,610
3	0	C_3^5 $\times$	C_2^{42} $\times$	C_0^1 $\times$	C_1^9	$= 10 \times 861 \times 1 \times 9 =$	77,490
2	1	C_2^5 $\times$	C_3^{42} $\times$	C_1^1 $\times$	C_0^9	$= 10 \times 11,480 \times 1 \times 1 =$	114,800
2	0	C_2^5 $\times$	C_3^{42} $\times$	C_0^1 $\times$	C_1^9	$= 10 \times 11,480 \times 1 \times 9 =$	1,033,200

A lottery of this type is played in the United Kingdom.

5/47 Plus 1/27

The lottery drawing machine (Machine A) contains forty-seven balls numbered from 1 to 47. Five balls are drawn without replacement. These are the machine A winning numbers.

A second lottery drawing machine (Machine B) contains twenty-seven balls numbered from 1 to 27. One ball is drawn without replacement. This is the machine B winning number.

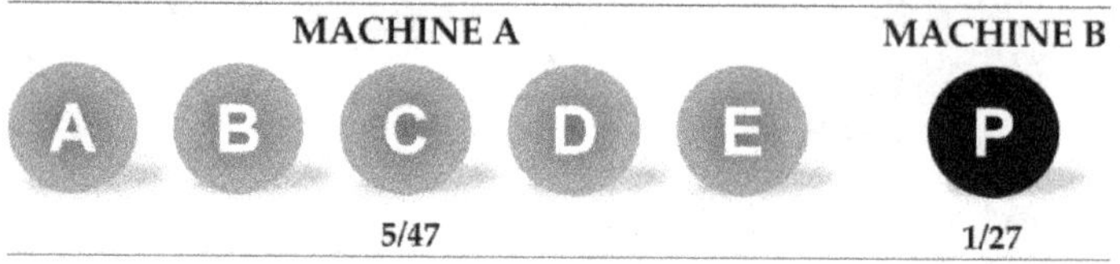

Prize Structure and Probabilities

5/47 plus 1/27

	Match			
	Machine A Winning Numbers	Machine B Winning Number	Favorable Outcomes	Probabilities
				1 in
DIV 1	5	1	1	41,416,353
DIV 2	5		26	1,592,937
DIV 3	4	1	210	197,221
DIV 4	4		5,460	7,585
DIV 5	3	1	8,610	4,810
DIV 6	3		223,860	185
DIV 7	2	1	114,800	361
DIV 8	1	1	559,650	74
DIV 9		1	850,668	49

Total Possible Outcomes

$$C(47,5) \times C(27,1) = 1,533,939 \times 27 = 41,416,353$$

Calculation of Favorable Outcomes

Match A	Match B	MACHINE A W		MACHINE B W		W = Winning numbers	
5	1	C_5^5	× C_0^{42}	× C_1^1	× C_0^{26}	= 1 x 1 x 1 x 1 =	1
5	0	C_5^5	× C_0^{42}	× C_0^1	× C_1^{26}	= 1 x 1 x 1 x 26 =	26
4	1	C_4^5	× C_1^{42}	× C_1^1	× C_0^{26}	= 5 x 42 x 1 x 1 =	210
4	0	C_4^5	× C_1^{42}	× C_0^1	× C_1^{26}	= 5 x 42 x 1 x 26 =	5,460
3	1	C_3^5	× C_2^{42}	× C_1^1	× C_0^{26}	= 10 x 861 x 1 x 1 =	8,610
3	0	C_3^5	× C_2^{42}	× C_0^1	× C_1^{26}	= 10 x 861 x 1 x 26 =	223,860
2	1	C_2^5	× C_3^{42}	× C_1^1	× C_0^{26}	= 10 x 11,480 x 1 x 1 =	114,800
1	1	C_1^5	× C_4^{42}	× C_1^1	× C_0^{26}	= 5 x 111,930 x 1 x 1 =	559,650
0	1	C_0^5	× C_5^{42}	× C_1^1	× C_0^{26}	= 1 x 850,668 x 1 x 1 =	850,668

A lottery of this type is played in California, USA.

5/48 Plus 1/18

The lottery drawing machine (Machine A) contains forty-eight balls numbered from 1 to 48. Five balls are drawn without replacement. These are the machine A winning numbers.

A second lottery ball drawing machine (Machine B) contains eighteen balls numbered from 1 to 18. One ball is drawn without replacement. This is the machine B winning number

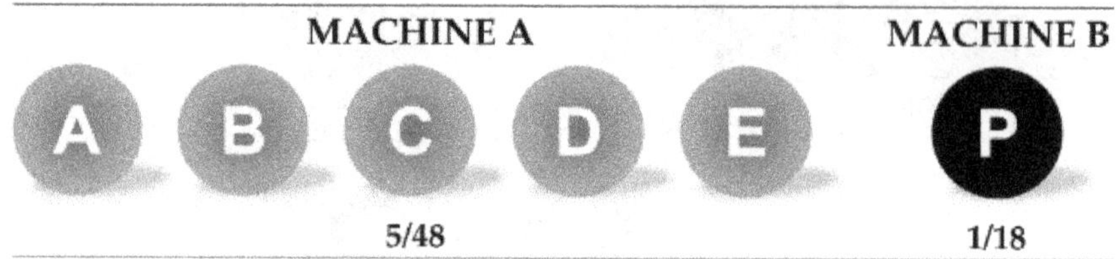

Prize Structure and Probabilities

5/48 plus 1/18

	Match Machine A Winning Numbers	Machine B Winning Number	Favorable Outcomes	Probabilities 1 in
DIV 1	5	1	1	30,821,472
DIV 2	5		17	1,813,028
DIV 3	4	1	215	143,356
DIV 4	4		3,655	8,433
DIV 5	3	1	9,030	3,413
DIV 6	3		153,510	201
DIV 7	2	1	123,410	250
DIV 8	2		2,097,970	15
DIV 9	1	1	617,050	50
DIV 10		1	962,598	32

Total Possible Outcomes

$$C(48,5) \times C(18,1) = 1{,}712{,}304 \times 18 = 30{,}821{,}472$$

Calculation of Favorable Outcomes

Match A	Match B	MACHINE A (W)		MACHINE B (W)		W = Winning numbers	
5	1	C_5^5	$\times\ C_0^{43}$	$\times\ C_1^1$	$\times\ C_0^{17}$	$= 1 \times 1 \times 1 \times 1 =$	1
5	0	C_5^5	$\times\ C_0^{43}$	$\times\ C_0^1$	$\times\ C_1^{17}$	$= 1 \times 1 \times 1 \times 17 =$	17
4	1	C_4^5	$\times\ C_1^{43}$	$\times\ C_1^1$	$\times\ C_0^{17}$	$= 5 \times 43 \times 1 \times 1 =$	215
4	0	C_4^5	$\times\ C_1^{43}$	$\times\ C_0^1$	$\times\ C_1^{17}$	$= 5 \times 43 \times 1 \times 17 =$	3,655
3	1	C_3^5	$\times\ C_2^{43}$	$\times\ C_1^1$	$\times\ C_0^{17}$	$= 10 \times 903 \times 1 \times 1 =$	9,030
3	0	C_3^5	$\times\ C_2^{43}$	$\times\ C_0^1$	$\times\ C_1^{17}$	$= 10 \times 903 \times 1 \times 17 =$	153,510
2	1	C_2^5	$\times\ C_3^{43}$	$\times\ C_1^1$	$\times\ C_0^{17}$	$= 10 \times 12{,}341 \times 1 \times 1 =$	123,410
2	0	C_2^5	$\times\ C_3^{43}$	$\times\ C_0^1$	$\times\ C_1^{17}$	$= 10 \times 12{,}341 \times 1 \times 17 =$	2,097,970
1	1	C_1^5	$\times\ C_4^{43}$	$\times\ C_1^1$	$\times\ C_0^{17}$	$= 5 \times 123{,}410 \times 1 \times 1 =$	617,050
0	1	C_0^5	$\times\ C_5^{43}$	$\times\ C_1^1$	$\times\ C_0^{17}$	$= 1 \times 962{,}598 \times 1 \times 1 =$	962,598

A lottery of this type is played in 22 states,
and the District of Columbia, of the USA.

5/49 Plus 1/10

The lottery drawing machine (Machine A) contains forty-nine balls numbered from 1 to 49. Five balls are drawn without replacement. These are the machine A winning numbers.

A second lottery drawing machine (Machine B) contains ten balls numbered from 1 to 10. One ball is drawn without replacement. This is the machine B winning number.

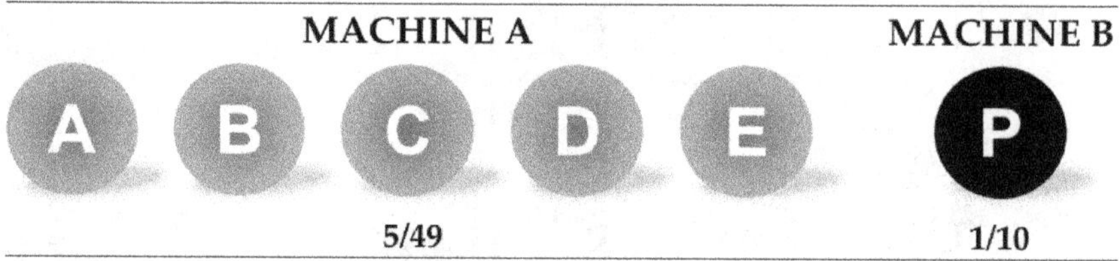

Prize Structure and Probabilities

5/49 plus 1/10

	Match			
	Machine A Winning Numbers	Machine B Winning Number	Favorable Outcomes	Probabilities
				1 in
DIV 1	5	1	1	19,068,840
DIV 2	5		9	2,118,760
DIV 3	4	1	220	86,677
DIV 4	4		1,980	9,631
DIV 5	3	1	9,460	2,016
DIV 6	3		85,140	224
DIV 7	2	1	132,440	144
DIV 8	2		1,191,960	16
DIV 9	1	1	678,755	28
DIV 10		1	1,086,008	18

Total Possible Outcomes

$$C(49,5) \times C(10,1) = 1{,}906{,}884 \times 10 = 19{,}068{,}840$$

Calculation of Favorable Outcomes

Match A	B	MACHINE A W		MACHINE B W		W = Winning numbers	
5	1	C_5^5	$\times\ C_0^{44}$	$\times\ C_1^1$	$\times\ C_0^9$	$= 1 \times 1 \times 1 \times 1 =$	1
5	0	C_5^5	$\times\ C_0^{44}$	$\times\ C_0^1$	$\times\ C_1^9$	$= 1 \times 1 \times 1 \times 9 =$	9
4	1	C_4^5	$\times\ C_1^{44}$	$\times\ C_1^1$	$\times\ C_0^9$	$= 5 \times 44 \times 1 \times 1 =$	220
4	0	C_4^5	$\times\ C_1^{44}$	$\times\ C_0^1$	$\times\ C_1^9$	$= 5 \times 44 \times 1 \times 9 =$	1,980
3	1	C_3^5	$\times\ C_2^{44}$	$\times\ C_1^1$	$\times\ C_0^9$	$= 10 \times 946 \times 1 \times 1 =$	9,460
3	0	C_3^5	$\times\ C_2^{44}$	$\times\ C_0^1$	$\times\ C_1^9$	$= 10 \times 946 \times 1 \times 9 =$	85,140
2	1	C_2^5	$\times\ C_3^{44}$	$\times\ C_1^1$	$\times\ C_0^9$	$= 10 \times 13{,}244 \times 1 \times 1 =$	132,440
2	0	C_2^5	$\times\ C_3^{44}$	$\times\ C_0^1$	$\times\ C_1^9$	$= 10 \times 13{,}244 \times 1 \times 9 =$	1,191,960
1	1	C_1^5	$\times\ C_4^{44}$	$\times\ C_1^1$	$\times\ C_0^9$	$= 5 \times 135{,}751 \times 1 \times 1 =$	678,755
0	1	C_0^5	$\times\ C_5^{44}$	$\times\ C_1^1$	$\times\ C_0^9$	$= 1 \times 1{,}086{,}008 \times 1 \times 1 =$	1,086,008

A lottery of this type is played in France.

5/49 Plus 1/13

The lottery drawing machine (Machine A) contains forty-nine balls numbered from 1 to 49. Five balls are drawn without replacement. These are the machine A winning numbers.

A second lottery drawing machine (Machine B) contains thirteen balls numbered from 1 to 13. One ball is drawn without replacement. This is the machine B winning number.

Prize Structure and Probabilities

5/49 plus 1/13

	Match		Favorable Outcomes	Probabilities
	Machine A Winning Numbers	Machine B Winning Number		
				1 in
DIV 1	5	1	1	24,789,492
DIV 2	5	0	12	2,065,791
DIV 3	4	0	2,640	9,390
DIV 4	3	0	113,520	218
DIV 5	2	0	1,589,280	16
DIV 6	0	1	1,086,008	23

See page 315 for a definition of this prize structure.

Total Possible Outcomes

$$C(49,5) \times C(13,1) = 1,906,884 \times 13 = 24,789,492$$

Calculation of Favorable Outcomes

Match A	Match B	MACHINE A		MACHINE B		W = Winning numbers	
		W		**W**			
5	1	C_5^5 ×	C_0^{44} ×	C_1^1 ×	C_0^{12}	= 1 x 1 x 1 x 1 =	1
5	0	C_5^5 ×	C_0^{44} ×	C_0^1 ×	C_1^{12}	= 1 x 1 x 1 x 12 =	12
4	0	C_4^5 ×	C_1^{44} ×	C_0^1 ×	C_1^{12}	= 5 x 44 x 1 x 12 =	2,640
3	0	C_3^5 ×	C_2^{44} ×	C_0^1 ×	C_1^{12}	= 10 x 946 x 1 x 12 =	113,520
2	0	C_2^5 ×	C_3^{44} ×	C_0^1 ×	C_1^{12}	= 10 x 13,244 x 1 x 12 =	1,589,280
0	1	C_0^5 ×	C_5^{44} ×	C_1^1 ×	C_0^{12}	= 1 x 1,086,008 x 1 x 1 =	1,086,008

A lottery of this type is played in Portugal.

5/50 Plus 2/10

The lottery drawing machine (Machine A) contains fifty balls numbered from 1 to 50. Five balls are drawn without replacement. These are the machine A winning numbers.

A second lottery drawing machine (Machine B) contains ten balls numbered from 1 to 10. Two balls are drawn without replacement. These are the machine B winning numbers.

Prize Structure and Probabilities

5/50 plus 2/10

	Match Machine A Winning Numbers	Machine B Winning Numbers	Favorable Outcomes	Probabilities
				1 in
DIV 1	5	2	1	95,344,200
DIV 2	5	1	16	5,959,013
DIV 3	4		28	3,405,150
DIV 4	4	2	225	423,752
DIV 5	3	1	3,600	26,485
DIV 6	3		6,300	15,134
DIV 7	3	2	9,900	9,631
DIV 8	2	1	158,400	602
DIV 9	2		277,200	344
DIV 10	2	2	141,900	672
DIV 11	2	1	2,270,400	42
DIV 12	1	2	744,975	128

Total Possible Outcomes

$$C(50,5) \times C(10,2) = 2{,}118{,}760 \times 45 = 95{,}344{,}200$$

Calculation of Favorable Outcomes

Match A	Match B	MACHINE A — W		MACHINE B — W		W = Winning numbers	
5	2	C_5^5	$\times\ C_0^{45}$	$\times\ C_2^2$	$\times\ C_0^8$	$= 1 \times 1 \times 1 \times 1 =$	1
5	1	C_5^5	$\times\ C_0^{45}$	$\times\ C_1^2$	$\times\ C_1^8$	$= 1 \times 1 \times 2 \times 8 =$	16
5	0	C_5^5	$\times\ C_0^{45}$	$\times\ C_0^2$	$\times\ C_2^8$	$= 1 \times 1 \times 1 \times 28 =$	28
4	2	C_4^5	$\times\ C_1^{45}$	$\times\ C_2^2$	$\times\ C_0^8$	$= 5 \times 45 \times 1 \times 1 =$	225
4	1	C_4^5	$\times\ C_1^{45}$	$\times\ C_1^2$	$\times\ C_1^8$	$= 5 \times 45 \times 2 \times 8 =$	3,600
4	0	C_4^5	$\times\ C_1^{45}$	$\times\ C_0^2$	$\times\ C_2^8$	$= 5 \times 45 \times 1 \times 28 =$	6,300
3	2	C_3^5	$\times\ C_2^{45}$	$\times\ C_2^2$	$\times\ C_0^8$	$= 10 \times 990 \times 1 \times 1 =$	9,900
3	1	C_3^5	$\times\ C_2^{45}$	$\times\ C_1^2$	$\times\ C_1^8$	$= 10 \times 990 \times 2 \times 8 =$	158,400
3	0	C_3^5	$\times\ C_2^{45}$	$\times\ C_0^2$	$\times\ C_2^8$	$= 10 \times 990 \times 1 \times 28 =$	277,200
2	2	C_2^5	$\times\ C_3^{45}$	$\times\ C_2^2$	$\times\ C_0^8$	$= 10 \times 14{,}190 \times 1 \times 1 =$	141,900
2	1	C_2^5	$\times\ C_3^{45}$	$\times\ C_1^2$	$\times\ C_1^8$	$= 10 \times 14{,}190 \times 2 \times 8 =$	2,270,400
1	2	C_1^5	$\times\ C_4^{45}$	$\times\ C_2^2$	$\times\ C_0^8$	$= 5 \times 148{,}995 \times 1 \times 1 =$	744,975

A lottery of this type was played in Europe until March 2022.

5/50 Plus 2/12

The lottery drawing machine (Machine A) contains fifty balls numbered from 1 to 50. Five balls are drawn without replacement. These are the machine A winning numbers. A second lottery drawing machine (Machine B) contains twelve balls numbered from 1 to 12. Two balls are drawn without replacement. These are the machine B winning numbers.

	MACHINE A					MACHINE B	
	A	B	C	D	E	P	Q
			5/50			2/12	

Prize Structure and Probabilities

5/50 plus 2/12

	Match		Favorable Outcomes	Probabilities
	Machine **A** Winning Numbers	Machine **B** Winning Numbers		1 in
DIV 1	5	2	1	139,838,160
DIV 2	5	1	20	6,991,908
DIV 3	5		45	3,107,515
DIV 4	4	2	225	621,503
DIV 5	4	1	4,500	31,075
DIV 6	4		10,125	13,811
DIV 7	3	2	9,900	14,125
DIV 8	3	1	198,000	706
DIV 9	3		445,500	314
DIV 10	2	2	141,900	985
DIV 11	2	1	2,838,000	49
DIV 12	2		6,385,500	22
DIV 13	1	2	744,975	188

Total Possible Outcomes

$$C(50,5) \times C(12,2) = 2{,}118{,}760 \times 66 = 139{,}838{,}160$$

Calculation of Favorable Outcomes

Match A	Match B	MACHINE A (W)		MACHINE B (W)		W = Winning numbers	
5	2	$C_5^5 \times$	$C_0^{45} \times$	$C_2^2 \times$	C_0^{10}	$= 1 \times 1 \times 1 \times 1 =$	1
5	1	$C_5^5 \times$	$C_0^{45} \times$	$C_1^2 \times$	C_1^{10}	$= 1 \times 1 \times 2 \times 10 =$	20
5	0	$C_5^5 \times$	$C_0^{45} \times$	$C_0^2 \times$	C_2^{10}	$= 1 \times 1 \times 1 \times 45 =$	45
4	2	$C_4^5 \times$	$C_1^{45} \times$	$C_2^2 \times$	C_0^{10}	$= 5 \times 45 \times 1 \times 1 =$	225
4	1	$C_4^5 \times$	$C_1^{45} \times$	$C_1^2 \times$	C_1^{10}	$= 5 \times 45 \times 2 \times 10 =$	4,500
4	0	$C_4^5 \times$	$C_1^{45} \times$	$C_0^2 \times$	C_2^{10}	$= 5 \times 45 \times 1 \times 45 =$	10,125
3	2	$C_3^5 \times$	$C_2^{45} \times$	$C_2^2 \times$	C_0^{10}	$= 10 \times 990 \times 1 \times 1 =$	9,900
3	1	$C_3^5 \times$	$C_2^{45} \times$	$C_1^2 \times$	C_1^{10}	$= 10 \times 990 \times 2 \times 10 =$	198,000
3	0	$C_3^5 \times$	$C_2^{45} \times$	$C_0^2 \times$	C_2^{10}	$= 10 \times 990 \times 1 \times 45 =$	445,500
2	2	$C_2^5 \times$	$C_3^{45} \times$	$C_2^2 \times$	C_0^{10}	$= 10 \times 14{,}190 \times 1 \times 1 =$	141,900
2	1	$C_2^5 \times$	$C_3^{45} \times$	$C_1^2 \times$	C_1^{10}	$= 10 \times 14{,}190 \times 2 \times 10 =$	2,838,000
2	0	$C_2^5 \times$	$C_3^{45} \times$	$C_0^2 \times$	C_2^{10}	$= 10 \times 14{,}190 \times 1 \times 45 =$	6,385,500
1	2	$C_1^5 \times$	$C_4^{45} \times$	$C_2^2 \times$	C_0^{10}	$= 5 \times 148{,}995 \times 1 \times 1 =$	744,975

> Two lotteries of this type are played
> in eighteen European countries.
> Their prize structures differ by only one category.

5/52 Plus 1/10

One lottery drawing machine (Machine A) contains fifty-two balls numbered from 1 to 52. Five balls are drawn without replacement. These are the machine A winning numbers.

A second lottery ball drawing machine (Machine B) contains ten balls numbered from 1 to 10. One ball is drawn without replacement. This is the machine B winning number.

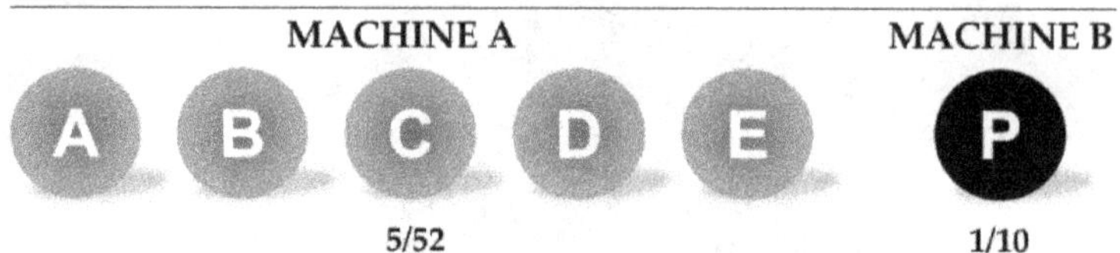

Prize Structure and Probabilities

5/52 plus 1/10

	Match			
	Machine **A** Winning Numbers	Machine **B** Winning Number	Favorable Outcomes	Probabilities
				1 in
DIV 1	5	1	1	25,989,600
DIV 2	5		9	2,887,733
DIV 3	4	1	235	110,594
DIV 4	4		2,115	12,288
DIV 5	3	1	10,810	2,404
DIV 6	3		97,290	267
DIV 7	2	1	162,150	160
DIV 8	1	1	891,825	29
DIV 9		1	1,533,939	17

Total Possible Outcomes

$$C(52,5) \times C(10,1) = 2{,}598{,}960 \times 10 = 25{,}989{,}600$$

Calculation of Favorable Outcomes

Match A	Match B	MACHINE A (W)		MACHINE B (W)		W = Winning numbers	
5	1	C_5^5 x C_0^{47}	x	C_1^1 x C_0^9	= 1 x 1 x 1 x 1 =	1
5	0	C_5^5 x C_0^{47}	x	C_0^1 x C_1^9	= 1 x 1 x 1 x 9 =	9
4	1	C_4^5 x C_1^{47}	x	C_1^1 x C_0^9	= 5 x 47 x 1 x 1 =	235
4	0	C_4^5 x C_1^{47}	x	C_0^1 x C_1^9	= 5 x 47 x 1 x 9 =	2,115
3	1	C_3^5 x C_2^{47}	x	C_1^1 x C_0^9	= 10 x 1,081 x 1 x 1 =	10,810
3	0	C_3^5 x C_2^{47}	x	C_0^1 x C_1^9	= 10 x 1,081 x 1 x 9 =	97,290
2	1	C_2^5 x C_3^{47}	x	C_1^1 x C_0^9	= 10 x 16,215 x 1 x 1 =	162,150
1	1	C_1^5 x C_4^{47}	x	C_1^1 x C_0^9	= 5 x 178,365 x 1 x 1 =	891,825
0	1	C_0^5 x C_5^{47}	x	C_1^1 x C_0^9	= 1 x 1,533,939 x 1 x 1 =	1,533,939

A lottery of this type is played in 13 states
of the United States of America.

5/54 Plus 1/10

One lottery drawing machine (Machine A) contains fifty-four balls numbered from 1 to 54. Five balls are drawn without replacement. These are the machine A winning numbers.

A second lottery ball drawing machine (Machine B) contains ten balls numbered from 1 to 10. One ball is drawn without replacement. This is the machine B winning number.

Prize Structure and Probabilities

5/54 plus 1/10

	Match			
	Machine A Winning Numbers	Machine B Winning Number	Favorable Outcomes	Probabilities
				1 in
DIV 1	5	1	1	31,625,100
DIV 2	5		9	3,513,900
DIV 3	4	1	245	129,082
DIV 4	4		2,205	14,342
DIV 5	3	1	11,760	2,689
DIV 6	3		105,840	299
DIV 7	2	1	184,240	172
DIV 8	2		1,658,160	19

Total Possible Outcomes

$$C(54,5) \times C(10,1) = 3,162,510 \times 10 = 31,625,100$$

Calculation of Favorable Outcomes

Match A	Match B	MACHINE A W		MACHINE B W		W= Winning numbers	
5	1	C_5^5 ×	C_0^{49} ×	C_1^1 ×	C_0^9	= 1 x 1 x 1 x 1 =	1
5	0	C_5^5 ×	C_0^{49} ×	C_0^1 ×	C_1^9	= 1 x 1 x 1 x 9 =	9
4	1	C_4^5 ×	C_1^{49} ×	C_1^1 ×	C_0^9	= 5 x 49 x 1 x 1 =	245
4	0	C_4^5 ×	C_1^{49} ×	C_0^1 ×	C_1^9	= 5 x 49 x 1 x 9 =	2,205
3	1	C_3^5 ×	C_2^{49} ×	C_1^1 ×	C_0^9	= 10 x 1,176 x 1 x 1 =	11,760
3	0	C_3^5 ×	C_2^{49} ×	C_0^1 ×	C_1^9	= 10 x 1,176 x 1 x 9 =	105,840
2	1	C_2^5 ×	C_3^{49} ×	C_1^1 ×	C_0^9	= 10 x 18,424 x 1 x 1 =	184,240
2	0	C_2^5 ×	C_3^{49} ×	C_0^1 ×	C_1^9	= 10 x 18,424 x 1 x 9 =	1,658,160

A lottery of this type is played in Spain.

5/60 Plus 1/4

One lottery drawing machine (Machine A) contains sixty balls numbered from 1 to 60. Five balls are drawn without replacement. These are the machine A winning numbers.

A second lottery ball drawing machine (Machine B) contains four balls numbered from 1 to 4. One ball is drawn without replacement. This is the machine B winning number.

Prize Structure and Probabilities

5/60 plus 1/4			
	Match		
	Machine A Winning Numbers	Machine B Winning Number	Favorable Outcomes	Probabilities

	Machine A Winning Numbers	Machine B Winning Number	Favorable Outcomes	Probabilities 1 in
DIV 1	5	1	1	21,846,048
DIV 2	5		3	7,282,016
DIV 3	4	1	275	79,440
DIV 4	4		825	26,480
DIV 5	3	1	14,850	1,471
DIV 6	3		44,550	490
DIV 7	2	1	262,350	83
DIV 8	2		787,050	28
DIV 9	1	1	1,705,275	13

Total Possible Outcomes

$$C(60,5) \times C(4,1) = 5,461,512 \times 4 = 21,846,048$$

Calculation of Favorable Outcomes

Match A	Match B	MACHINE A — W		MACHINE B — W		W = Winning numbers	
5	1	C_5^5	$\times\ C_0^{55}$	C_1^1	$\times\ C_0^3$	$= 1 \times 1 \times 1 \times 1 =$	1
5	0	C_5^5	$\times\ C_0^{55}$	C_0^1	$\times\ C_1^3$	$= 1 \times 1 \times 1 \times 3 =$	3
4	1	C_4^5	$\times\ C_1^{55}$	C_1^1	$\times\ C_0^3$	$= 5 \times 55 \times 1 \times 1 =$	275
4	0	C_4^5	$\times\ C_1^{55}$	C_0^1	$\times\ C_1^3$	$= 5 \times 55 \times 1 \times 3 =$	825
3	1	C_3^5	$\times\ C_2^{55}$	C_1^1	$\times\ C_0^3$	$= 10 \times 1{,}485 \times 1 \times 1 =$	14,850
3	0	C_3^5	$\times\ C_2^{55}$	C_0^1	$\times\ C_1^3$	$= 10 \times 1{,}485 \times 1 \times 3 =$	44,550
2	1	C_2^5	$\times\ C_3^{55}$	C_1^1	$\times\ C_0^3$	$= 10 \times 26{,}235 \times 1 \times 1 =$	262,350
2	0	C_2^5	$\times\ C_3^{55}$	C_0^1	$\times\ C_1^3$	$= 10 \times 26{,}235 \times 1 \times 3 =$	787,050
1	1	C_1^5	$\times\ C_4^{55}$	C_1^1	$\times\ C_0^3$	$= 5 \times 341{,}055 \times 1 \times 1 =$	1,705,275

A lottery of this type is played in ten states of the United States of America, namely Florida, Georgia, Indiana, Maryland, Missouri, New Jersey, New York, Pennsylvania, Tennessee, and Virginia.

5/69 Plus 1/26

One lottery drawing machine (Machine A) contains sixty-nine balls numbered from 1 to 69. Five balls are drawn without replacement. These are the machine A winning numbers.

A second lottery drawing machine (Machine B) contains twenty-six balls numbered from 1 to 26. One ball is drawn without replacement. This is the machine B winning number.

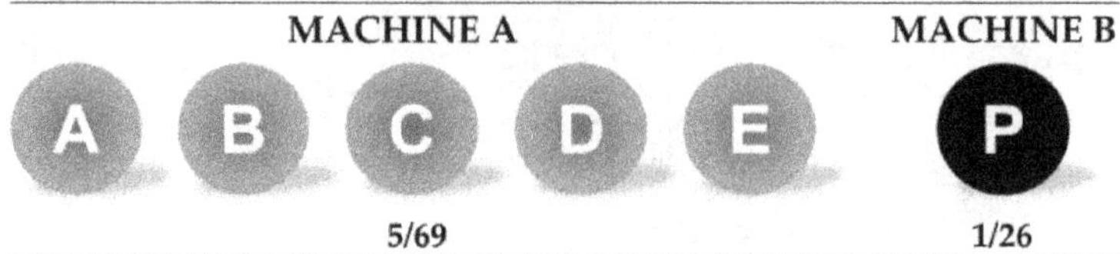

Prize Structure and Probabilities

	5/69 plus 1/26			
	Match			
	Machine A Winning Numbers	Machine B Winning Number	Favorable Outcomes	Probabilities
				1 in
DIV 1	5	1	1	292,201,338
DIV 2	5		25	11,688,054
DIV 3	4	1	320	913,129
DIV 4	4		8,000	36,525
DIV 5	3	1	20,160	14,494
DIV 6	3		504,000	580
DIV 7	2	1	416,640	701
DIV 8	1	1	3,176,880	92
DIV 9		1	7,624,512	38

Total Possible Outcomes

$$C(69,5) \times C(26,1) = 11,238,513 \times 26 = 292,201,338$$

Calculation of Favorable Outcomes

Match A	Match B	MACHINE A (W)		MACHINE B (W)		W = Winning numbers	
5	1	C_5^5 ×	C_0^{64} ×	C_1^1 ×	C_0^{25}	= 1 x 1 x 1 x 1 =	1
5	0	C_5^5 ×	C_0^{64} ×	C_0^1 ×	C_1^{25}	= 1 x 1 x 1 x 25 =	25
4	1	C_4^5 ×	C_1^{64} ×	C_1^1 ×	C_0^{25}	= 5 x 64 x 1 x 1 =	320
4	0	C_4^5 ×	C_1^{64} ×	C_0^1 ×	C_1^{25}	= 5 x 64 x 1 x 25 =	8,000
3	1	C_3^5 ×	C_2^{64} ×	C_1^1 ×	C_0^{25}	= 10 x 2,016 x 1 x 1 =	20,160
3	0	C_3^5 ×	C_2^{64} ×	C_0^1 ×	C_1^{25}	= 10 x 2,016 x 1 x 25 =	504,000
2	1	C_2^5 ×	C_3^{64} ×	C_1^1 ×	C_0^{25}	= 10 x 41,664 x 1 x 1 =	416,640
1	1	C_1^5 ×	C_4^{64} ×	C_1^1 ×	C_0^{25}	= 5 x 635,376 x 1 x 1 =	3,176,880
0	1	C_0^5 ×	C_5^{64} ×	C_1^1 ×	C_0^{25}	= 1 x 7,624,512 x 1 x 1 =	7,624,512

A lottery of this type is played in 45 states of the United States of America, the District of Columbia, Puerto Rico, and the U.S. Virgin Islands.

5/70 Plus 1/25

One lottery drawing machine (Machine A) contains seventy balls numbered from 1 to 70. Five balls are drawn without replacement. These are the machine A winning numbers.

A second lottery drawing machine (Machine B) contains twenty-five balls numbered from 1 to 25. One ball is drawn without replacement. This is the machine B winning number.

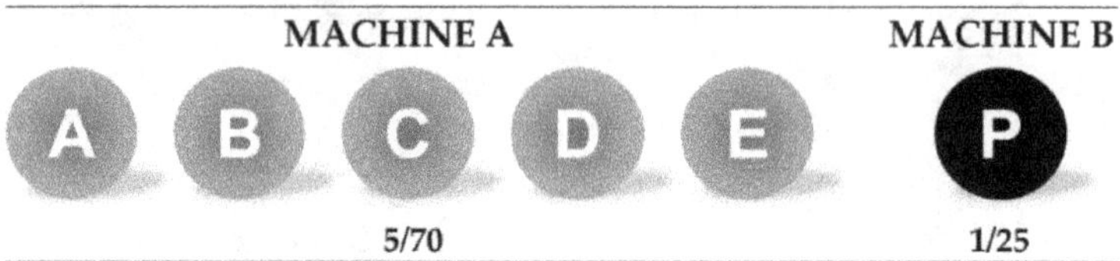

Prize Structure and Probabilities

	5/70 plus 1/25			
	Match			
	Machine A Winning Numbers	Machine B Winning Number	Favorable Outcomes	Probabilities
				1 in
DIV 1	5	1	1	302,575,350
DIV 2	5		24	12,607,306
DIV 3	4	1	325	931,001
DIV 4	4		7,800	38,792
DIV 5	3	1	20,800	14,547
DIV 6	3		499,200	606
DIV 7	2	1	436,800	693
DIV 8	1	1	3,385,200	89
DIV 9		1	8,259,888	37

Total Possible Outcomes

$$C(70,5) \times C(25,1) = 12,103,104 \times 25 = 302,575,350$$

Calculation of Favorable Outcomes

Match		MACHINE A		MACHINE B		W = Winning numbers
A	**B**	**W**		**W**		
5	1	$C_5^5 \times C_0^{65}$		$C_1^1 \times C_0^{24}$		$= 1 \times 1 \times 1 \times 1 =$ 1
5	0	$C_5^5 \times C_0^{65}$		$C_0^1 \times C_1^{24}$		$= 1 \times 1 \times 1 \times 24 =$ 24
4	1	$C_4^5 \times C_1^{65}$		$C_1^1 \times C_0^{24}$		$= 5 \times 65 \times 1 \times 1 =$ 325
4	0	$C_4^5 \times C_1^{65}$		$C_0^1 \times C_1^{24}$		$= 5 \times 65 \times 1 \times 24 =$ 7,800
3	1	$C_3^5 \times C_2^{65}$		$C_1^1 \times C_0^{24}$		$= 10 \times 2,080 \times 1 \times 1 =$ 20,800
3	0	$C_3^5 \times C_2^{65}$		$C_0^1 \times C_1^{24}$		$= 10 \times 2,080 \times 1 \times 24 =$ 499,200
2	1	$C_2^5 \times C_3^{65}$		$C_1^1 \times C_0^{24}$		$= 10 \times 43,680 \times 1 \times 1 =$ 436,800
1	1	$C_1^5 \times C_4^{65}$		$C_1^1 \times C_0^{24}$		$= 5 \times 677,040 \times 1 \times 1 =$ 3,385,200
0	1	$C_0^5 \times C_5^{65}$		$C_1^1 \times C_0^{24}$		$= 1 \times 8,259,888 \times 1 \times 1 =$ 8,259,888

A lottery of this type is played in 45 states of the
United States of America, the District of Columbia,
and the U.S. Virgin Islands.

6/40 (1 supplementary number) Plus 1/10

One lottery drawing machine (Machine A) contains forty balls numbered from 1 to 40. Six balls are drawn without replacement. These are the machine A winning numbers. One more ball is drawn without replacement. This is the supplementary number.

A second lottery ball drawing machine (Machine B) contains ten balls numbered from 1 to 10. One ball is drawn without replacement. This is the machine B winning number.

MACHINE A	MACHINE B
A B C D E F S	P
6/40 (1 supplementary number)	1/10

Prize Structure and Probabilities

6/40 (1 supplementary number) Plus 1/10

| | Match | | | | |
	Machine A		Machine B	Favorable Outcomes	Probabilities
	Winning Numbers	Suppl. Number	Winning Number		1 in
DIV 1	6		1	1	38,383,800
DIV 2	5	1	1	6	6,397,300
DIV 3	5		1	198	193,858
DIV 4	4	1	1	495	77,543
DIV 5	4		1	7,920	4,846
DIV 6	3	1	1	10,560	3,635
DIV 7	3		1	109,120	352

Total Possible Outcomes

$$C(40,6) \times C(10,1) = 3,838,380 \times 10 = 38,383,800$$

LOTTERIES

Calculation of Favorable Outcomes					
W = Winning numbers (6)		**S** = Suppl. numbers (1)		**R** = Remaining numbers (33)	
MATCH		**MACHINE A**			
W	**S**	**W**	**S**	**R**	
DIV 1	6	0	$C_6^6 \times C_0^1 \times C_0^{33} = 1 \times 1 \times 1 =$	1
DIV 2	5	1	$C_5^6 \times C_1^1 \times C_0^{33} = 6 \times 1 \times 1 =$	6
DIV 3	5	0	$C_5^6 \times C_0^1 \times C_1^{33} = 6 \times 1 \times 33 =$	198
DIV 4	4	1	$C_4^6 \times C_1^1 \times C_1^{33} = 15 \times 1 \times 33 =$	495
DIV 5	4	0	$C_4^6 \times C_0^1 \times C_2^{33} = 15 \times 1 \times 528 =$	7,920
DIV 6	3	1	$C_3^6 \times C_1^1 \times C_2^{33} = 20 \times 1 \times 528 =$	10,560
DIV 7	3	0	$C_3^6 \times C_0^1 \times C_3^{33} = 20 \times 1 \times 5,456 =$	109,120

	Match	MACHINE B	
	W	**W**	**R**
DIV 1	1	0	$C_1^1 \times C_0^9 = 1 \times 1 =$	1
DIV 2	1	0	$C_1^1 \times C_0^9 = 1 \times 1 =$	1
DIV 3	1	0	$C_1^1 \times C_0^9 = 1 \times 1 =$	1
DIV 4	1	0	$C_1^1 \times C_0^9 = 1 \times 1 =$	1
DIV 5	1	0	$C_1^1 \times C_0^9 = 1 \times 1 =$	1
DIV 6	1	0	$C_1^1 \times C_0^9 = 1 \times 1 =$	1
DIV 7	1	0	$C_1^1 \times C_0^9 = 1 \times 1 =$	1

Because the winning number from the machine B must be matched in all divisions, this factor is always 1, and therefore the total number of outcomes for each division is the same as shown on the table above.

See page 316.

A lottery of this type is played in New Zealand.

6/42 Plus 1/6

One lottery drawing machine (Machine A) contains forty-two balls numbered from 1 to 42. Six balls are drawn without replacement. These are the machine A winning numbers.

A second lottery drawing machine (Machine B) contains six balls numbered from 1 to 6. One ball is drawn without replacement. This is the machine B winning number.

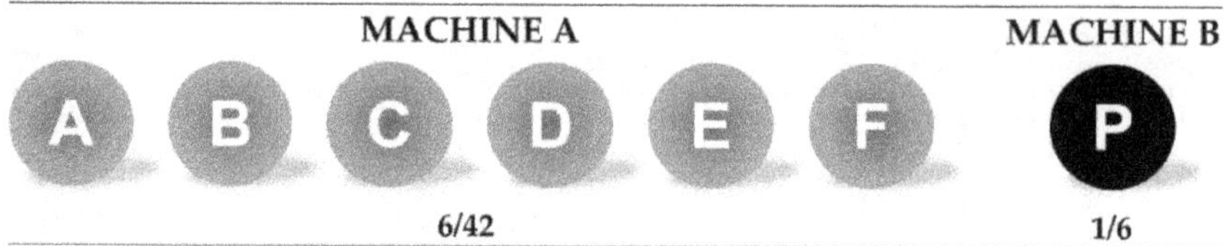

Prize Structure and Probabilities

6/42 plus 1/6

	Match			
	Machine **A** Winning Numbers	Machine **B** Winning Number	Favorable Outcomes	Probabilities
				1 in
DIV 1	6	1	1	31,474,716
DIV 2	6		5	6,294,943
DIV 3	5	1	216	145,716
DIV 4	5		1,080	29,143
DIV 5	4	1	9,450	3,331
DIV 6	4		47,250	666
DIV 7	3	1	142,800	220
DIV 8	3		714,000	44

Total Possible Outcomes

$$C(42,6) \times C(6,1) = 5,245,786 \times 6 = 31,474,716$$

Calculation of Favorable Outcomes

Match A	Match B	MACHINE A — W		MACHINE B — W		W = Winning numbers	
6	1	C_6^6 ×	C_0^{36} ×	C_1^1 ×	C_0^5	= 1 x 1 x 1 x 1 =	1
6	0	C_6^6 ×	C_0^{36} ×	C_0^1 ×	C_1^5	= 1 x 1 x 1 x 5 =	5
5	1	C_5^6 ×	C_1^{36} ×	C_1^1 ×	C_0^5	= 6 x 36 x 1 x 1 =	216
5	0	C_5^6 ×	C_1^{36} ×	C_0^1 ×	C_1^5	= 6 x 36 x 1 x 5 =	1,080
4	1	C_4^6 ×	C_2^{36} ×	C_1^1 ×	C_0^5	= 15 x 630 x 1 x 1 =	9,450
4	0	C_4^6 ×	C_2^{36} ×	C_0^1 ×	C_1^5	= 15 x 630 x 1 x 5 =	47,250
3	1	C_3^6 ×	C_3^{36} ×	C_1^1 ×	C_0^5	= 20 x 7,140 x 1 x 1 =	142,800
3	0	C_3^6 ×	C_3^{36} ×	C_0^1 ×	C_1^5	= 20 x 7,140 x 1 x 5 =	714,000

A lottery of this type is played in Switzerland.

6/42 Plus 1/10

One lottery drawing machine (Machine A) contains forty-two balls numbered from 1 to 42. Six balls are drawn without replacement. These are the machine A winning numbers.

A second lottery drawing machine (Machine B) contains ten balls numbered from 1 to 10. One ball is drawn without replacement. This is the machine B winning number.

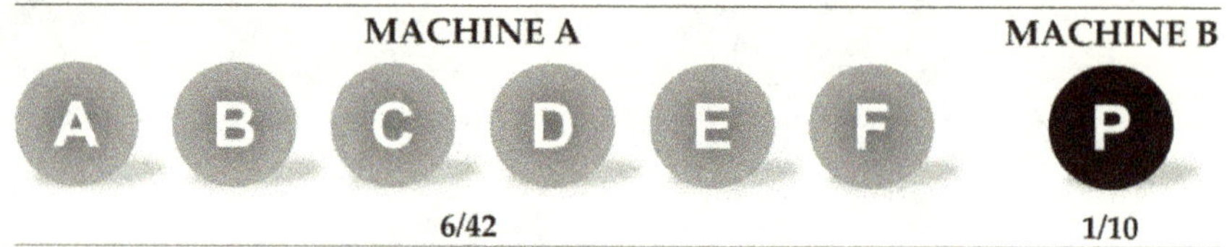

Prize Structure and Probabilities

6/42 plus 1/10

	Match			
	Machine A Winning Numbers	Machine B Winning Number	Favorable Outcomes	Probabilities
				1 in
DIV 1	6	1	1	52,457,860
DIV 2	6		9	5,828,651
DIV 3	5	1	216	242,860
DIV 4	5		1,944	26,984
DIV 5	4	1	9,450	5,551
DIV 6	4		85,050	617
DIV 7	3	1	142,800	367
DIV 8	3		1,285,200	41

Total Possible Outcomes

$$C(42,6) \times C(10,1) = 5{,}245{,}786 \times 10 = 52{,}457{,}860$$

Calculation of Favorable Outcomes

Match A	Match B	MACHINE A W		MACHINE B W		Winning numbers	
6	1	C_6^6 ×	C_0^{36} ×	C_1^1 ×	C_0^9	= 1 x 1 x 1 x 1 =	1
6	0	C_6^6 ×	C_0^{36} ×	C_0^1 ×	C_1^9	= 1 x 1 x 1 x 9 =	9
5	1	C_5^6 ×	C_1^{36} ×	C_1^1 ×	C_0^9	= 6 x 36 x 1 x 1 =	216
5	0	C_5^6 ×	C_1^{36} ×	C_0^1 ×	C_1^9	= 6 x 36 x 1 x 9 =	1,944
4	1	C_4^6 ×	C_2^{36} ×	C_1^1 ×	C_0^9	= 15 x 630 x 1 x 1 =	9,450
4	0	C_4^6 ×	C_2^{36} ×	C_0^1 ×	C_1^9	= 15 x 630 x 1 x 9 =	85,050
3	1	C_3^6 ×	C_3^{36} ×	C_1^1 ×	C_0^9	= 20 x 7,140 x 1 x 1 =	142,800
3	0	C_3^6 ×	C_3^{36} ×	C_0^1 ×	C_1^9	= 20 x 7,140 x 1 x 9 =	1,285,200

A lottery of this type is played in Ukraine.

6/42 Plus 2/10

A lottery drawing machine (Machine A) contains forty-two balls numbered from 0 to 41. Six balls are drawn without replacement. These are the machine A winning numbers.

A second lottery drawing machine (Machine B) contains ten balls numbered from 0 to 9. Two balls are drawn without replacement. These are the machine B winning numbers.

Prize Structure and Probabilities

6/42 plus 2/10

	Match			
	Machine A Winning Numbers	Machine B Winning Numbers	Favorable Outcomes	Probabilities
				1 in
DIV 1	6	2	1	236,060,370
DIV 2	6	1	16	14,753,773
DIV 3	6		28	8,430,727
DIV 4	5	2	216	1,092,872
DIV 5	5	1	3,456	68,305
DIV 6	5		6,048	39,031

Total Possible Outcomes

$$C(42,6) \times C(10,2) = 5{,}245{,}786 \times 45 = 236{,}060{,}370$$

Calculation of Favorable Outcomes

Match A	Match B	MACHINE A — W		MACHINE B — W		Winning numbers	
6	2	C_6^6 ×	C_0^{36} ×	C_2^2 ×	C_0^8	= 1 x 1 x 1 x 1 =	1
6	1	C_6^6 ×	C_0^{36} ×	C_1^2 ×	C_1^8	= 1 x 1 x 2 x 8 =	16
6	0	C_6^6 ×	C_0^{36} ×	C_0^2 ×	C_2^8	= 1 x 1 x 1 x 28 =	28
5	2	C_5^6 ×	C_1^{36} ×	C_2^2 ×	C_0^8	= 6 x 36 x 1 x 1 =	216
5	1	C_5^6 ×	C_1^{36} ×	C_1^2 ×	C_1^8	= 6 x 36 x 2 x 8 =	3,456
5	0	C_5^6 ×	C_1^{36} ×	C_0^2 ×	C_2^8	= 6 x 36 x 1 x 28 =	6,048

A lottery of this type is played in Argentina.

6/48 Plus 1/5

One lottery drawing machine (Machine A) contains forty-eight balls numbered from 1 to 48. Six balls are drawn without replacement. These are the machine A winning numbers.

A second lottery drawing machine (Machine B) contains five balls numbered from 1 to 5. One ball is drawn without replacement. This is the machine B winning number.

Prize Structure and Probabilities

6/48 plus 1/5

	Match			
	Machine A Winning Numbers	Machine B Winning Number	Favorable Outcomes	Probabilities
				1 in
DIV 1	6	1	1	61,357,560
DIV 2	6		4	15,339,390
DIV 3	5	1	252	243,482
DIV 4	5		1,008	60,871
DIV 5	4	1	12,915	4,751
DIV 6	4		51,660	1,188
DIV 7	3	1	229,600	267
DIV 8	3		918,400	67
DIV 9	2	1	1,678,950	37
DIV 10	2		6,715,800	9

Total Possible Outcomes

$$C(48,6) \times C(5,1) = 12{,}271{,}512 \times 5 = 61{,}357{,}560$$

Calculation of Favorable Outcomes

Match A	Match B	MACHINE A (W)		MACHINE B (W)		W = Winning numbers	
6	1	C^6_6	$\times\ C^{42}_0$	$\times\ C^1_1$	$\times\ C^4_0$	$= 1 \times 1 \times 1 \times 1 =$	1
6	0	C^6_6	$\times\ C^{42}_0$	$\times\ C^1_0$	$\times\ C^4_1$	$= 1 \times 1 \times 1 \times 4 =$	4
5	1	C^6_5	$\times\ C^{42}_1$	$\times\ C^1_1$	$\times\ C^4_0$	$= 6 \times 42 \times 1 \times 1 =$	252
5	0	C^6_5	$\times\ C^{42}_1$	$\times\ C^1_0$	$\times\ C^4_1$	$= 6 \times 42 \times 1 \times 4 =$	1,008
4	1	C^6_4	$\times\ C^{42}_2$	$\times\ C^1_1$	$\times\ C^4_0$	$= 15 \times 861 \times 1 \times 1 =$	12,915
4	0	C^6_4	$\times\ C^{42}_2$	$\times\ C^1_0$	$\times\ C^4_1$	$= 15 \times 861 \times 1 \times 4 =$	51,660
3	1	C^6_3	$\times\ C^{42}_3$	$\times\ C^1_1$	$\times\ C^4_0$	$= 20 \times 11{,}480 \times 1 \times 1 =$	229,600
3	0	C^6_3	$\times\ C^{42}_3$	$\times\ C^1_0$	$\times\ C^4_1$	$= 20 \times 11{,}480 \times 1 \times 4 =$	918,400
2	1	C^6_2	$\times\ C^{42}_4$	$\times\ C^1_1$	$\times\ C^4_0$	$= 15 \times 111{,}930 \times 1 \times 1 =$	1,678,950
2	0	C^6_2	$\times\ C^{42}_4$	$\times\ C^1_0$	$\times\ C^4_1$	$= 15 \times 111{,}930 \times 1 \times 4 =$	6,715,800

A lottery of this type is played in Belgium, Denmark, Estonia, Finland, Iceland, Latvia, Lithuania, Norway, Slovenia, and Sweden.

6/49 Plus 1/10

One lottery drawing machine (Machine A) contains forty-nine balls numbered from 1 to 49. Six balls are drawn without replacement. These are the machine A winning numbers.

A second lottery ball drawing machine (Machine B) contains ten balls numbered from 1 to 10. One ball is drawn without replacement. This is the machine B winning number.

Prize Structure and Probabilities

6/49 plus 1/10

	Match			
	Machine **A** Winning Numbers	Machine **B** Winning Number	Favorable Outcomes	Probabilities
				1 in
DIV 1	6	1	1	139,838,160
DIV 2	6		9	15,537,573
DIV 3	5	1	258	542,008
DIV 4	5		2,322	60,223
DIV 5	4	1	13,545	10,324
DIV 6	4		121,905	1,147
DIV 7	3	1	246,820	567
DIV 8	3		2,221,380	63
DIV 9	2	1	1,851,150	76

Total Possible Outcomes

C(49,6) x C(10,1) = 13,983,816 x 10 = 139,838,160

Calculation of Favorable Outcomes

Match A	B	MACHINE A W		MACHINE B W		W = Winning numbers	
6	1	C_6^6 ×	C_0^{43} ×	C_1^1 ×	C_0^9	= 1 x 1 x 1 x 1 =	1
6	0	C_6^6 ×	C_0^{43} ×	C_0^1 ×	C_1^9	= 1 x 1 x 1 x 9 =	9
5	1	C_5^6 ×	C_1^{43} ×	C_1^1 ×	C_0^9	= 6 x 43 x 1 x 1 =	258
5	0	C_5^6 ×	C_1^{43} ×	C_0^1 ×	C_1^9	= 6 x 43 x 1 x 9 =	2,322
4	1	C_4^6 ×	C_2^{43} ×	C_1^1 ×	C_0^9	= 15 x 903 x 1 x 1 =	13,545
4	0	C_4^6 ×	C_2^{43} ×	C_0^1 ×	C_1^9	= 15 x 903 x 1 x 9 =	121,905
3	1	C_3^6 ×	C_3^{43} ×	C_1^1 ×	C_0^9	= 20 x 12,341 x 1 x 1 =	246,820
3	0	C_3^6 ×	C_3^{43} ×	C_0^1 ×	C_1^9	= 20 x 12,341 x 1 x 9 =	2,221,380
2	1	C_2^6 ×	C_4^{43} ×	C_1^1 ×	C_0^9	= 15 x 123,410 x 1 x 1 =	1,851,150

A lottery of this type is played in Germany.

6/49 (1 supplementary number) Plus 1/10

One lottery drawing machine (Machine A) contains forty-nine balls numbered from 1 to 49. Six balls are drawn without replacement. These are the machine A winning numbers. One more ball is drawn without replacement. This is the supplementary number.

A second lottery ball drawing machine (Machine B) contains ten balls numbered from 1 to 10. One ball is drawn without replacement. This is the machine B winning number.

	MACHINE A	MACHINE B
A B C D E F **S**		**P**
6/49 (1 supplementary number)		1/10

Prize Structure and Probabilities

6/49 (1 supplementary) plus 1/10

	Match				
	Machine A		Machine B	Favorable	Probabilities
	Winning Numbers	Suppl. Number	Winning Number	Outcomes	
					1 in
1st Prize	6		1	1	139,838,160
DIV 1	6			9	15,537,573
DIV 2	5	1		60	2,330,636
DIV 3	5			2,520	55,491
DIV 4	4			135,450	1,032
DIV 5	3			2,468,200	57
Last prize			1	1	10

A lottery of this type is played in Spain.

Calculation of Favorable Outcomes

MACHINE A			MACHINE B				Outcomes	Outcomes	Probabilities 1 in	
W	S		W		W	S W				
$_6C_6 \times {}_1C_0 \times {}_{42}C_0 \times {}_1C_1 \times {}_9C_0 = 1 \times 1 \times 1 \times 1 \times 1 =$							1	1	139,838,160	1st PRIZE
$_6C_6 \times {}_1C_0 \times {}_{42}C_0 \times {}_1C_0 \times {}_9C_1 = 1 \times 1 \times 1 \times 1 \times 9 =$							9	9	15,537,573	DIV 1
$_6C_5 \times {}_1C_1 \times {}_{42}C_0 \times {}_1C_1 \times {}_9C_0 = 6 \times 1 \times 1 \times 1 \times 1 =$							6	60	2,330,636	DIV 2
$_6C_5 \times {}_1C_1 \times {}_{42}C_0 \times {}_1C_0 \times {}_9C_1 = 6 \times 1 \times 1 \times 1 \times 9 =$							54			
$_6C_5 \times {}_1C_0 \times {}_{42}C_1 \times {}_1C_1 \times {}_9C_0 = 6 \times 1 \times 42 \times 1 \times 1 =$							252	2,520	55,491	DIV 3
$_6C_5 \times {}_1C_0 \times {}_{42}C_1 \times {}_1C_0 \times {}_9C_1 = 6 \times 1 \times 42 \times 1 \times 9 =$							2,268			
$_6C_4 \times {}_1C_1 \times {}_{42}C_1 \times {}_1C_1 \times {}_9C_0 = 15 \times 1 \times 42 \times 1 \times 1 =$							630	135,450	1,032	DIV 4
$_6C_4 \times {}_1C_1 \times {}_{42}C_1 \times {}_1C_0 \times {}_9C_1 = 15 \times 1 \times 42 \times 1 \times 9 =$							5,670			
$_6C_4 \times {}_1C_0 \times {}_{42}C_2 \times {}_1C_1 \times {}_9C_0 = 15 \times 1 \times 861 \times 1 \times 1 =$							12,915			
$_6C_4 \times {}_1C_0 \times {}_{42}C_2 \times {}_1C_0 \times {}_9C_1 = 15 \times 1 \times 861 \times 1 \times 9 =$							116,235			
$_6C_3 \times {}_1C_1 \times {}_{42}C_2 \times {}_1C_1 \times {}_9C_0 = 20 \times 1 \times 861 \times 1 \times 1 =$							17,220	2,468,200	57	DIV 5
$_6C_3 \times {}_1C_1 \times {}_{42}C_2 \times {}_1C_0 \times {}_9C_1 = 20 \times 1 \times 861 \times 1 \times 9 =$							154,980			
$_6C_3 \times {}_1C_0 \times {}_{42}C_3 \times {}_1C_1 \times {}_9C_0 = 20 \times 1 \times 11,480 \times 1 \times 1 =$							229,600			
$_6C_3 \times {}_1C_0 \times {}_{42}C_3 \times {}_1C_0 \times {}_9C_1 = 20 \times 1 \times 11,480 \times 1 \times 9 =$							2,066,400			

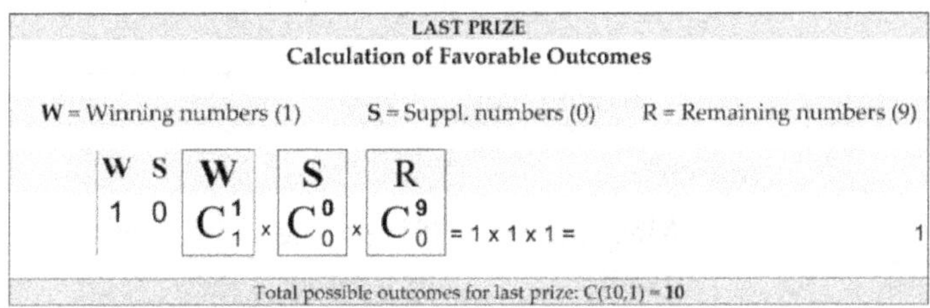

LAST PRIZE
Calculation of Favorable Outcomes

W = Winning numbers (1)   S = Suppl. numbers (0)   R = Remaining numbers (9)

$$\underset{1}{W} \;\; \underset{0}{S} \qquad {}_1C_1^{1} \times {}_0C_0^{0} \times {}_0C_0^{9} = 1 \times 1 \times 1 = \qquad 1$$

Total possible outcomes for last prize: C(10,1) = **10**

To better understand these calculations, see page 317, where you can compare the full tables for both <u>6/49 (1 supplementary number) Plus 1/10</u>, and <u>6/49 Plus 1/10</u> (no supplementary number).

6/50 plus 1/25

One lottery drawing machine (Machine A) contains fifty balls numbered from 1 to 50. Six balls are drawn without replacement. These are the machine A winning numbers.

A second lottery drawing machine (Machine B) contains twenty-five balls numbered from 1 to 25. One ball is drawn without replacement. This is the machine B winning number.

Prize Structure and Probabilities

6/50 plus 1/25

	Match			
	Machine **A** Winning Numbers	Machine **B** Winning Number	Favorable Outcomes	Probabilities
				1 in
DIV 1	6		1	15,890,700
DIV 2	5	1	264	1,504,801
DIV 3	5		264	60,192
DIV 4	4	1	14,190	27,996
DIV 5	4		14,190	1,120
DIV 6	3	1	264,880	1,500
DIV 7	3		264,880	60
DIV 8	2	1	2,036,265	195
DIV 9	2		2,036,265	8
DIV 10	1	1	6,516,048	61
DIV 11		1	7,059,052	56

In this lottery the player has the option of <u>not choosing</u> any number to play from the machine B. This decision transforms the prize structure in two different alternatives:

Alternative 1	Alternative 2
Total possible outcomes for **Machine A** combinations	Total possible outcomes for **Machine A plus Machine B** combinations
$C(50,6) = $ **15,890,700**	$C(50,6) \times C(25,1) = 15{,}890{,}700 \times 25 = $ **397,267,500**

Calculation of Favorable Outcomes

Match A	B	MACHINE A W		MACHINE B W		W = Winning numbers
6	0	C_6^6	$\times\ C_0^{44}$			$= 1 \times 1 = $ 1
5	1	C_5^6	$\times\ C_1^{44}$	$\times\ C_1^1$	$\times\ C_0^{24}$	$= 6 \times 44 \times 1 \times 1 = $ 264
5	0	C_5^6	$\times\ C_1^{44}$			$= 6 \times 44 = $ 264
4	1	C_4^6	$\times\ C_2^{44}$	$\times\ C_1^1$	$\times\ C_0^{24}$	$= 15 \times 946 \times 1 \times 1 = $ 14,190
4	0	C_4^6	$\times\ C_2^{44}$			$= 15 \times 946 = $ 14,190
3	1	C_3^6	$\times\ C_3^{44}$	$\times\ C_1^1$	$\times\ C_0^{24}$	$= 20 \times 13{,}244 \times 1 \times 1 = $ 264,880
3	0	C_3^6	$\times\ C_3^{44}$			$= 20 \times 13{,}244 = $ 264,880
2	1	C_2^6	$\times\ C_4^{44}$	$\times\ C_1^1$	$\times\ C_0^{24}$	$= 15 \times 135{,}751 \times 1 \times 1 = $ 2,036,265
2	0	C_2^6	$\times\ C_4^{44}$			$= 15 \times 135{,}751 = $ 2,036,265
1	1	C_1^6	$\times\ C_5^{44}$	$\times\ C_1^1$	$\times\ C_0^{24}$	$= 6 \times 1{,}086{,}008 \times 1 \times 1 = $ 6,516,048
0	1	C_0^6	$\times\ C_6^{44}$			$= 1 \times 7{,}059{,}052 = $ 7,059,052

A lottery of this type is played in Illinois, USA.

6/90 (1 supplementary number) Plus 1/90

One lottery drawing machine (Machine A) contains ninety balls numbered from 1 to 90. Six balls are drawn without replacement. These are the machine A winning numbers. One more ball is drawn without replacement. This is the supplementary number.

A second lottery drawing machine (Machine B) contains ninety balls numbered from 1 to 90. One ball is drawn without replacement. This is the machine B winning number.

MACHINE A							MACHINE B
A	B	C	D	E	F	S	P
		6/90 (1 supplementary number)					1/90

Prize Structure and Probabilities

6/90 (1 supplementary) plus 1/90

	Match				
	Machine A		Machine B	Favorable Outcomes	Probabilities
	Winning Numbers	Suppl. Number	Winning Number		
					1 in
DIV 1	6		1	1	56,035,316,700
DIV 2	5	1	1	6	9,339,219,450
DIV 3	5		1	498	112,520,716
DIV 4	4		1	52,290	1,071,626
DIV 5	3		1	1,905,680	29,404
DIV 6	2		1	28,942,515	1,936
DIV 7	1		1	185,232,096	303
DIV 8			1	406,481,544	138

Total Possible Outcomes

$$C(90,6) \times C(90,1) = 622,614,630 \times 90 = 56,035,316,700$$

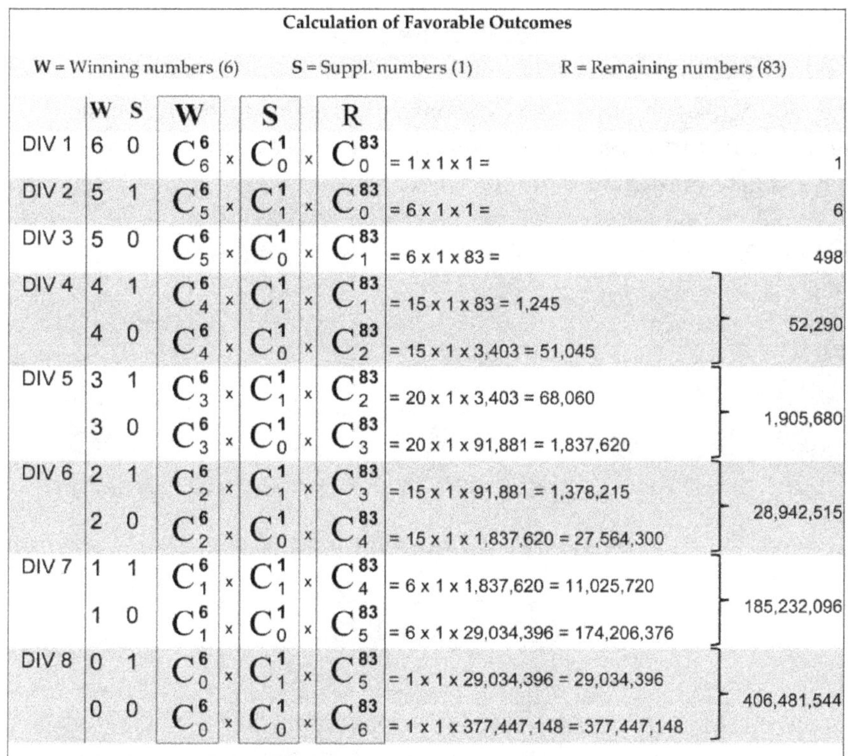

Calculation of Favorable Outcomes

W = Winning numbers (6)   S = Suppl. numbers (1)   R = Remaining numbers (83)

	W	S	W		S		R		
DIV 1	6	0	C_6^6	x	C_0^1	x	C_0^{83}	= 1 x 1 x 1 =	1
DIV 2	5	1	C_5^6	x	C_1^1	x	C_0^{83}	= 6 x 1 x 1 =	6
DIV 3	5	0	C_5^6	x	C_0^1	x	C_1^{83}	= 6 x 1 x 83 =	498
DIV 4	4	1	C_4^6	x	C_1^1	x	C_1^{83}	= 15 x 1 x 83 = 1,245	52,290
	4	0	C_4^6	x	C_0^1	x	C_2^{83}	= 15 x 1 x 3,403 = 51,045	
DIV 5	3	1	C_3^6	x	C_1^1	x	C_2^{83}	= 20 x 1 x 3,403 = 68,060	1,905,680
	3	0	C_3^6	x	C_0^1	x	C_3^{83}	= 20 x 1 x 91,881 = 1,837,620	
DIV 6	2	1	C_2^6	x	C_1^1	x	C_3^{83}	= 15 x 1 x 91,881 = 1,378,215	28,942,515
	2	0	C_2^6	x	C_0^1	x	C_4^{83}	= 15 x 1 x 1,837,620 = 27,564,300	
DIV 7	1	1	C_1^6	x	C_1^1	x	C_4^{83}	= 6 x 1 x 1,837,620 = 11,025,720	185,232,096
	1	0	C_1^6	x	C_0^1	x	C_5^{83}	= 6 x 1 x 29,034,396 = 174,206,376	
DIV 8	0	1	C_0^6	x	C_1^1	x	C_5^{83}	= 1 x 1 x 29,034,396 = 29,034,396	406,481,544
	0	0	C_0^6	x	C_0^1	x	C_6^{83}	= 1 x 1 x 377,447,148 = 377,447,148	

Because the winning number from the machine B must be matched in all divisions, this factor is always 1, and therefore the total number of outcomes for each division is the same as shown on the table above.

A lottery of this type is played in Italy.

7/35 Plus 1/20

One lottery drawing machine (Machine A) contains thirty-five balls numbered from 1 to 35. Seven balls are drawn without replacement. These are the machine A winning numbers.

A second lottery drawing machine (Machine B) contains twenty balls numbered from 1 to 20. One ball is drawn without replacement. This is the machine B winning number.

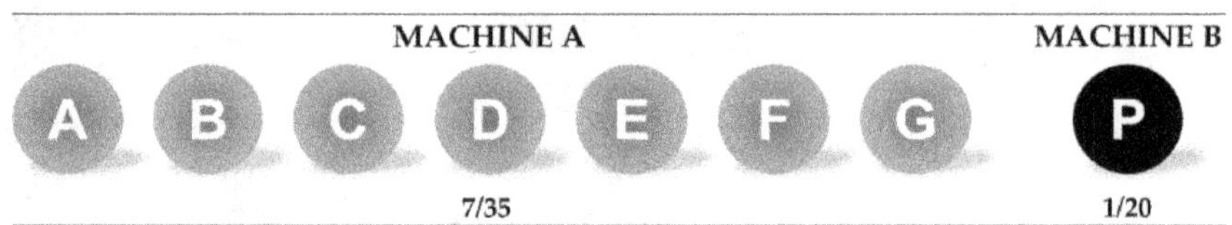

Prize Structure and Probabilities

7/35 plus 1/20

	Match			
	Machine A Winning Numbers	Machine B Winning Number	Favorable Outcomes	Probabilities
				1 in
DIV 1	7	1	1	134,490,400
DIV 2	7		19	7,078,442
DIV 3	6	1	196	686,176
DIV 4	6		3,724	36,115
DIV 5	5	1	7,938	16,943
DIV 6	5		150,822	892
DIV 7	4	1	114,660	1,173
DIV 8	3	1	716,625	188
DIV 9	2	1	2,063,880	65

Total Possible Outcomes

$$C(35,7) \times C(20,1) = 6,724,520 \times 20 = 134,490,400$$

Calculation of Favorable Outcomes					
Match A B	MACHINE A		MACHINE B		W = Winning numbers
	W		W		
7 1	C_7^7 ×	C_0^{28} ×	C_1^1 ×	C_0^{19}	= 1 x 1 x 1 x 1 = 1
7 0	C_7^7 ×	C_0^{28} ×	C_0^1 ×	C_1^{19}	= 1 x 1 x 1 x 19 = 19
6 1	C_6^7 ×	C_1^{28} ×	C_1^1 ×	C_0^{19}	= 7 x 28 x 1 x 1 = 196
6 0	C_6^7 ×	C_1^{28} ×	C_0^1 ×	C_1^{19}	= 7 x 28 x 1 x 19 = 3,724
5 1	C_5^7 ×	C_2^{28} ×	C_1^1 ×	C_0^{19}	= 21 x 378 x 1 x 1 = 7,938
5 0	C_5^7 ×	C_2^{28} ×	C_0^1 ×	C_1^{19}	= 21 x 378 x 1 x 19 = 150,822
4 1	C_4^7 ×	C_3^{28} ×	C_1^1 ×	C_0^{19}	= 35 x 3,276 x 1 x 1 = 114,660
3 1	C_3^7 ×	C_4^{28} ×	C_1^1 ×	C_0^{19}	= 35 x 20,475 x 1 x 1 = 716,625
2 1	C_2^7 ×	C_5^{28} ×	C_1^1 ×	C_0^{19}	= 21 x 98,280 x 1 x 1 = 2,063,880

A lottery of this type is played in Australia.

CHAPTER 7

SELECTED TOPICS

7.1 Which Lottery is the Easiest to Win?

It sounds reasonable that playing the lottery game offering the highest probability of winning a prize would be the smartest choice. The probabilities of the different prize divisions of the huge number of lotteries in the world can range from those so low that they are nearly impossible to win, to the ones so high that the prize you can win is negligible.

The table below shows the probabilities of winning the first division prize for the lotteries studied in this book (EO: Exact Order):

LOTTERY	PROBABILITIES			
	1 GAME	2 GAMES	4 GAMES	8 GAMES
	1 in	1 in	1 in	1 in
3/10 EO	1,000	500	250	125
4/10 EO	10,000	5,000	2,500	1,250
5/28	98,280	49,140	24,570	12,285
5/10 EO	100,000	50,000	25,000	12,500
2/26 Plus 2/26	105,625	52,813	26,406	13,203
5/31	169,911	84,956	42,478	21,239
5/32	201,376	100,688	50,344	25,172
5/35	324,632	162,316	81,158	40,579
5/36	376,992	188,496	94,248	47,124
5/39	575,757	287,879	143,940	71,970
6/30	593,775	296,888	148,444	74,222
11/22	705,432	352,716	176,358	88,179
5/41	749,398	374,699	87,350	93,675
5/42	850,668	425,334	212,667	106,334
6/10 EO	1,000,000	500,000	250,000	125,000
5/45	1,221,759	610,880	305,440	152,720
4/35 Plus 1/25	1,309,000	654,500	327,250	163,625
5/35 Plus 1/5	1,623,160	811,580	405,790	202,895
4/35 Plus 1/35	1,832,600	916,300	458,150	229,075

4/40 EO	2,193,360	1,096,680	548,340	274,170
4/45 EO	3,575,880	1,787,940	893,970	446,985
6/40	3,838,380	1,919,190	959,595	479,798
5/41 Plus 1/6	4,496,388	2,248,194	1,124,097	562,049
6/41	4,496,388	2,248,194	1,124,097	562,049
5/32 Plus 1/25	5,034,400	2,517,200	1,258,600	629,300
6/42	5,245,786	2,622,893	1,311,447	655,723
7/34	5,379,616	2,689,808	1,344,904	672,452
6/43	6,096,454	3,048,227	1,524,114	762,057
7/35	6,724,520	3,362,260	1,681,130	840,565
6/44	7,059,052	3,529,526	1,764,763	882,382
5/39 Plus 1/14	8,060,598	4,030,299	2,015,150	1,007,575
6/45	8,145,060	4,072,530	2,036,265	1,018,133
6/46	9,366,819	4,683,410	2,341,705	1,170,852
7/37	10,295,472	5,147,736	2,573,868	1,286,934
6/47	10,737,573	5,368,787	2,684,393	1,342,197
6/49	13,983,816	6,991,908	3,495,954	1,747,977
5/47 Plus 1/10	15,339,390	7,669,695	3,834,848	1,917,424
5/43 Plus 1/16	15,401,568	7,700,784	3,850,392	1,925,196
7/40	18,643,560	9,321,780	4,660,890	2,330,445
5/49 Plus 1/10	19,068,840	9,534,420	4,767,210	2,383,605
6/52	20,358,520	10,179,260	5,089,630	2,544,815
5/60 Plus 1/4	21,846,048	10,923,024	5,461,512	2,730,756
6/53	22,957,480	11,478,740	5,739,370	2,869,685
5/80	24,040,016	12,020,008	6,010,004	3,005,002
5/49 Plus 1/13	24,789,492	12,394,746	6,197,373	3,098,687
6/54	25,827,165	12,913,583	6,456,791	3,228,396
5/52 Plus 1/10	25,989,600	12,994,800	6,497,400	3,248,700
6/55	28,989,675	14,494,838	7,247,419	3,623,709
5/48 Plus 1/18	30,821,472	15,410,736	7,705,368	3,852,684
6/42 Plus 1/6	31,474,716	15,737,358	7,868,679	3,934,340
5/54 Plus 1/10	31,625,100	15,812,550	7,906,275	3,953,138
6/56	32,468,436	16,234,218	8,117,109	4,058,555
7/44	38,320,568	19,160,284	9,580,142	4,790,071
6/58	40,475,358	20,237,679	10,118,840	5,059,420
5/47 Plus 1/27	41,416,353	20,708,177	10,354,088	5,177,044

5/90	43,949,268	21,974,634	10,987,317	5,493,659
6/59	45,057,474	22,528,737	11,264,369	5,632,184
7/45	45,379,620	22,689,810	11,344,905	5,672,453
6/60	50,063,860	25,031,930	12,515,965	6,257,983
6/42 Plus 1/10	52,457,860	26,228,930	13,114,465	6,557,233
6/48 Plus 1/5	61,357,560	30,678,780	15,339,390	7,669,695
7/47	62,891,499	31,445,750	15,722,875	7,861,437
5/50 Plus 2/10	95,344,200	47,672,100	23,836,050	11,918,025
7/50	99,884,400	49,942,200	24,971,100	12,485,550
7/35 Plus 1/20	134,490,400	67,245,200	33,622,600	16,811,300
5/50 Plus 2/12	139,838,160	69,919,080	34,959,540	17,479,770
6/49 Plus 1/10	139,838,160	69,919,080	34,959,540	17,479,770
5/69 Plus 1/26	292,201,338	146,100,669	73,050,335	36,525,167
5/70 Plus 1/25	302,575,350	151,287,675	75,643,838	37,821,919
6/90	622,614,630	311,307,315	155,653,658	77,826,829
6/90 Plus 1/90	56,035,316,700	28,017,658,350	14,008,829,175	7,004,414,588

7.2 The Fallacy of the Most Frequent Numbers

It is quite common to see lottery statistics in most official lottery websites. The information is generally presented in the form of tables, charts, and graphs. This information shows how often each number has been drawn, the most common lottery winning numbers, the most often picked numbers, the least often picked numbers, and those numbers which are most "overdue". Other tables show the number of days since each ball dropped out of the lottery machine as a winning number.

Searching the internet will result in a plethora of information on lotto number frequency tables, lotto number frequency graphs, most common lottery numbers, most common pairs, most common triplets, most common consecutive pairs, most common consecutive triplets, and so on.

But the truth is that the lottery number frequency has no meaning at all when working out the probabilities of matching the winning numbers. The phrase "lottery balls have no memory" has been said a thousand times, but people still do not believe it.

The following is a mathematical demonstration of the fact that the most frequent lottery number speculations are no more than a myth fuelled by the publication of frequency tables and graphs in the official lottery websites.

Frequency Distribution Tables

For any set of all possible combinations $C(n,r)$ a frequency distribution table can be made as follows:

- The rows represent the numbers 1 to n in ascending order
- The columns represent the numbers 1 to r in ascending order
- The values in the cells are the frequencies for each number in each position

In like manner, for any lottery without replacement and matrix m/M, a frequency distribution table can be made as follows:

- The rows represent the numbers 1 to M in ascending order
- The columns represent the numbers 1 to m in ascending order
- The values in the cells are the frequencies for each number in each position

For the purpose of this demonstration two types of tables are taken into consideration:

- ✓ Theoretical Frequency Distribution Tables
- ✓ Partial Frequency Distribution Tables

A theoretical frequency distribution table includes the frequency of each number in each position when all possible combinations are considered.

A partial frequency distribution table includes the frequencies of each number in each position when certain quantity, but <u>not all</u>, of the possible combinations are considered.

<u>Theoretical Frequency Distribution Tables</u>

These tables have the following characteristics:

- ✓ The sum of each column is the total number of possible combinations.
- ✓ The sum of each row is the total frequency of each number.

An example of a theoretical frequency distribution table is shown as Table 7.1A; it shows the values for lotto 6/45—8,145,060 combinations. The sum of each column represents the total number of possible combinations, which is

$$C(45,6) = 8,145,060$$

The sum of each row represents the total frequency of each number, which is 1,086,008. The total frequency of each number can be calculated as follows:

Because each combination has six numbers, there are

$$8,145,060 \times 6 = 48,870,360 \text{ numbers in the whole system}$$

And because there are 45 different numbers:

$$\text{Frequency for every number} = 48,870,360/45 = 1,086,008$$

TABLE 7.1A: Theoretical frequency distribution table for lotto 6/45 — 8,145,060 combinations.

	A	B	C	D	E	F	
1	1,086,008						1,086,008
2	962,598	123,410					1,086,008
3	850,668	223,860	11,480				1,086,008
4	749,398	303,810	31,980	820			1,086,008
5	658,008	365,560	59,280	3,120	40		1,086,008
6	575,757	411,255	91,390	7,410	195	1	1,086,008
7	501,942	442,890	126,540	14,060	570	6	1,086,008
8	435,897	462,315	163,170	23,310	1,295	21	1,086,008
9	376,992	471,240	199,920	35,280	2,520	56	1,086,008
10	324,632	471,240	235,620	49,980	4,410	126	1,086,008
11	278,256	463,760	269,280	67,320	7,140	252	1,086,008
12	237,336	450,120	300,080	87,120	10,890	462	1,086,008
13	201,376	431,520	327,360	109,120	15,840	792	1,086,008
14	169,911	409,045	350,610	132,990	22,165	1,287	1,086,008
15	142,506	383,670	369,460	158,340	30,030	2,002	1,086,008
16	118,755	356,265	383,670	184,730	39,585	3,003	1,086,008
17	98,280	327,600	393,120	211,680	50,960	4,368	1,086,008
18	80,730	298,350	397,800	238,680	64,260	6,188	1,086,008
19	65,780	269,100	397,800	265,200	79,560	8,568	1,086,008
20	53,130	240,350	393,300	290,700	96,900	11,628	1,086,008
21	42,504	212,520	384,560	314,640	116,280	15,504	1,086,008
22	33,649	185,955	371,910	336,490	137,655	20,349	1,086,008
23	26,334	160,930	355,740	355,740	160,930	26,334	1,086,008
24	20,349	137,655	336,490	371,910	185,955	33,649	1,086,008
25	15,504	116,280	314,640	384,560	212,520	42,504	1,086,008
26	11,628	96,900	290,700	393,300	240,350	53,130	1,086,008
27	8,568	79,560	265,200	397,800	269,100	65,780	1,086,008
28	6,188	64,260	238,680	397,800	298,350	80,730	1,086,008
29	4,368	50,960	211,680	393,120	327,600	98,280	1,086,008
30	3,003	39,585	184,730	383,670	356,265	118,755	1,086,008
31	2,002	30,030	158,340	369,460	383,670	142,506	1,086,008
32	1,287	22,165	132,990	350,610	409,045	169,911	1,086,008
33	792	15,840	109,120	327,360	431,520	201,376	1,086,008
34	462	10,890	87,120	300,080	450,120	237,336	1,086,008
35	252	7,140	67,320	269,280	463,760	278,256	1,086,008
36	126	4,410	49,980	235,620	471,240	324,632	1,086,008
37	56	2,520	35,280	199,920	471,240	376,992	1,086,008
38	21	1,295	23,310	163,170	462,315	435,897	1,086,008
39	6	570	14,060	126,540	442,890	501,942	1,086,008
40	1	195	7,410	91,390	411,255	575,757	1,086,008
41		40	3,120	59,280	365,560	658,008	1,086,008
42			820	31,980	303,810	749,398	1,086,008
43				11,480	223,860	850,668	1,086,008
44					123,410	962,598	1,086,008
45						1,086,008	1,086,008
	8,145,060	8,145,060	8,145,060	8,145,060	8,145,060	8,145,060	48,870,360	1,086,008

TABLE 7.1B: Theoretical frequency distribution table
for Lotto 6/45—8,145,060 combinations,
with frequencies expressed as fractions of the total.

	A	B	C	D	E	F	
1	0.133						0.133
2	0.118	0.015					0.133
3	0.104	0.027	0.001				0.133
4	0.092	0.037	0.004	0.000			0.133
5	0.081	0.045	0.007	0.000	0.000		0.133
6	0.071	0.050	0.011	0.001	0.000	0.000	0.133
7	0.062	0.054	0.016	0.002	0.000	0.000	0.133
8	0.054	0.057	0.020	0.003	0.000	0.000	0.133
9	0.046	0.058	0.025	0.004	0.000	0.000	0.133
10	0.040	0.058	0.029	0.006	0.001	0.000	0.133
11	0.034	0.057	0.033	0.008	0.001	0.000	0.133
12	0.029	0.055	0.037	0.011	0.001	0.000	0.133
13	0.025	0.053	0.040	0.013	0.002	0.000	0.133
14	0.021	0.050	0.043	0.016	0.003	0.000	0.133
15	0.017	0.047	0.045	0.019	0.004	0.000	0.133
16	0.015	0.044	0.047	0.023	0.005	0.000	0.133
17	0.012	0.040	0.048	0.026	0.006	0.001	0.133
18	0.010	0.037	0.049	0.029	0.008	0.001	0.133
19	0.008	0.033	0.049	0.033	0.010	0.001	0.133
20	0.007	0.030	0.048	0.036	0.012	0.001	0.133
21	0.005	0.026	0.047	0.039	0.014	0.002	0.133
22	0.004	0.023	0.046	0.041	0.017	0.002	0.133
23	0.003	0.020	0.044	0.044	0.020	0.003	0.133
24	0.002	0.017	0.041	0.046	0.023	0.004	0.133
25	0.002	0.014	0.039	0.047	0.026	0.005	0.133
26	0.001	0.012	0.036	0.048	0.030	0.007	0.133
27	0.001	0.010	0.033	0.049	0.033	0.008	0.133
28	0.001	0.008	0.029	0.049	0.037	0.010	0.133
29	0.001	0.006	0.026	0.048	0.040	0.012	0.133
30	0.000	0.005	0.023	0.047	0.044	0.015	0.133
31	0.000	0.004	0.019	0.045	0.047	0.017	0.133
32	0.000	0.003	0.016	0.043	0.050	0.021	0.133
33	0.000	0.002	0.013	0.040	0.053	0.025	0.133
34	0.000	0.001	0.011	0.037	0.055	0.029	0.133
35	0.000	0.001	0.008	0.033	0.057	0.034	0.133
36	0.000	0.001	0.006	0.029	0.058	0.040	0.133
37	0.000	0.000	0.004	0.025	0.058	0.046	0.133
38	0.000	0.000	0.003	0.020	0.057	0.054	0.133
39	0.000	0.000	0.002	0.016	0.054	0.062	0.133
40	0.000	0.000	0.001	0.011	0.050	0.071	0.133
41		0.000	0.000	0.007	0.045	0.081	0.133
42			0.000	0.004	0.037	0.092	0.133
43				0.001	0.027	0.104	0.133
44					0.015	0.118	0.133
45						0.133	0.133
	1.000	1.000	1.000	1.000	1.000	1.000	6.000	0.133

To abbreviate the table the quantities can be expressed as a fraction of the total. For example, the fraction of the total for the frequency of each number is

$$1,086,008/8,145,060 = 0.133$$

Table 7.1B shows the theoretical frequency distribution table for lotto 6/45—8,145,060 combinations, with the frequencies expressed as fractions of the total.

A second example of a theoretical frequency distribution table is shown as table 7.2A; it displays the values for lotto 5/39—575,757 combinations. The sum of each column represents the total number of possible combinations, which is

$$C(39,5) = 575,757$$

The sum of each row represents the frequency of each number, which is 73,815. The total frequency of each number can be calculated as follows:
Because each combination has five numbers, there are

$$575,757 \times 5 = 2,878,785 \text{ numbers in the whole system}$$

And because there are 39 different numbers:

$$\text{Frequency for every number} = 2,878,785/39 = 73,815$$

To abbreviate the table the quantities can be expressed as a fraction of the total. For example, the fraction of the total for the frequency of each number is

$$73,815/575,757 = 0.128$$

Table 7.2B shows the theoretical frequency distribution table for lotto 5/39—575,757 combinations, with the frequencies expressed as fractions of the total.

In conclusion, the total frequency for each number when all possible combinations of $C(n,r)$ are evaluated is invariable. Each number has the same total frequency.

Theoretical frequency distribution tables can be made for any m/M lottery, and the properties described above are applicable to all of them.

Note that the frequencies of each number change according to its position in the table, but the sum is invariable. For example, in lotto 6/45 the number 2 has a frequency of 962,598 in position A, and a frequency of

123,410 in position B. It could be said that playing number 2 in position A gives a better chance for that number, but never forget that the number of combinations following number 2 in columns C, D, E, and F is enormous!

TABLE 7.2A: Theoretical frequency distribution table for lotto 5/39 — 575,757 combinations.

	A	B	C	D	E	
1	73,815					73,815
2	66,045	7,770				73,815
3	58,905	14,280	630			73,815
4	52,360	19,635	1,785	35		73,815
5	46,376	23,936	3,366	136	1	73,815
6	40,920	27,280	5,280	330	5	73,815
7	35,960	29,760	7,440	640	15	73,815
8	31,465	31,465	9,765	1,085	35	73,815
9	27,405	32,480	12,180	1,680	70	73,815
10	23,751	32,886	14,616	2,436	126	73,815
11	20,475	32,760	17,010	3,360	210	73,815
12	17,550	32,175	19,305	4,455	330	73,815
13	14,950	31,200	21,450	5,720	495	73,815
14	12,650	29,900	23,400	7,150	715	73,815
15	10,626	28,336	25,116	8,736	1,001	73,815
16	8,855	26,565	26,565	10,465	1,365	73,815
17	7,315	24,640	27,720	12,320	1,820	73,815
18	5,985	22,610	28,560	14,280	2,380	73,815
19	4,845	20,520	29,070	16,320	3,060	73,815
20	3,876	18,411	29,241	18,411	3,876	73,815
21	3,060	16,320	29,070	20,520	4,845	73,815
22	2,380	14,280	28,560	22,610	5,985	73,815
23	1,820	12,320	27,720	24,640	7,315	73,815
24	1,365	10,465	26,565	26,565	8,855	73,815
25	1,001	8,736	25,116	28,336	10,626	73,815
26	715	7,150	23,400	29,900	12,650	73,815
27	495	5,720	21,450	31,200	14,950	73,815
28	330	4,455	19,305	32,175	17,550	73,815
29	210	3,360	17,010	32,760	20,475	73,815
30	126	2,436	14,616	32,886	23,751	73,815
31	70	1,680	12,180	32,480	27,405	73,815
32	35	1,085	9,765	31,465	31,465	73,815
33	15	640	7,440	29,760	35,960	73,815
34	5	330	5,280	27,280	40,920	73,815
35	1	136	3,366	23,936	46,376	73,815
36		35	1,785	19,635	52,360	73,815
37			630	14,280	58,905	73,815
38				7,770	66,045	73,815
39					73,815	73,815
	575,757	575,757	575,757	575,757	575,757	2,878,785	73,815

TABLE 7.2B: Theoretical frequency distribution table
for lotto 5/39 — 575,757 combinations,
with frequencies expressed as fractions of the total.

	A	B	C	D	E	
1	0.128					0.128
2	0.115	0.013				0.128
3	0.102	0.025	0.001			0.128
4	0.091	0.034	0.003	0.000		0.128
5	0.081	0.042	0.006	0.000	0.000	0.128
6	0.071	0.047	0.009	0.001	0.000	0.128
7	0.062	0.052	0.013	0.001	0.000	0.128
8	0.055	0.055	0.017	0.002	0.000	0.128
9	0.048	0.056	0.021	0.003	0.000	0.128
10	0.041	0.057	0.025	0.004	0.000	0.128
11	0.036	0.057	0.030	0.006	0.000	0.128
12	0.030	0.056	0.034	0.008	0.001	0.128
13	0.026	0.054	0.037	0.010	0.001	0.128
14	0.022	0.052	0.041	0.012	0.001	0.128
15	0.018	0.049	0.044	0.015	0.002	0.128
16	0.015	0.046	0.046	0.018	0.002	0.128
17	0.013	0.043	0.048	0.021	0.003	0.128
18	0.010	0.039	0.050	0.025	0.004	0.128
19	0.008	0.036	0.050	0.028	0.005	0.128
20	0.007	0.032	0.051	0.032	0.007	0.128
21	0.005	0.028	0.050	0.036	0.008	0.128
22	0.004	0.025	0.050	0.039	0.010	0.128
23	0.003	0.021	0.048	0.043	0.013	0.128
24	0.002	0.018	0.046	0.046	0.015	0.128
25	0.002	0.015	0.044	0.049	0.018	0.128
26	0.001	0.012	0.041	0.052	0.022	0.128
27	0.001	0.010	0.037	0.054	0.026	0.128
28	0.001	0.008	0.034	0.056	0.030	0.128
29	0.000	0.006	0.030	0.057	0.036	0.128
30	0.000	0.004	0.025	0.057	0.041	0.128
31	0.000	0.003	0.021	0.056	0.048	0.128
32	0.000	0.002	0.017	0.055	0.055	0.128
33	0.000	0.001	0.013	0.052	0.062	0.128
34	0.000	0.001	0.009	0.047	0.071	0.128
35	0.000	0.000	0.006	0.042	0.081	0.128
36		0.000	0.003	0.034	0.091	0.128
37			0.001	0.025	0.102	0.128
38				0.013	0.115	0.128
39					0.128	0.128
	1.000	1.000	1.000	1.000	1.000	5.000	0.128

Let us examine the following two hypothetical situations:

For Lotto 6/45: If there were a daily draw and each winning combination were unique every time, it would take approximately 22,396 years to obtain all 8,145,060 possible combinations.

This calculation assumes that a draw occurs every day without any skipped days.

For Lotto 5/39: If there were a daily draw and each winning combination were unique every time, it would take approximately 1,580 years to obtain all 575,757 possible combinations. Again, this calculation assumes a daily draw without any skipped days.

Lotteries are actually subsets of the whole system $C(n,r)$. Lotto m/M is a subset of $C(M,m)$.

Lotto 6/45 is a subset of $C(45,6)$, and lotto 5/39 is a subset of $C(39,5)$.

In other words, all results in the history of a lottery are one fraction of the total number of possible combinations. How large the fraction is depending on the timespan of the lottery.

The partial frequency distribution tables will allow us to compare the frequency of each number in one lottery to that in the theoretical system.

Partial Frequency Distribution Tables

These tables have the following characteristics:

- ✓ The sum of each column is the number of combinations under analysis.
- ✓ The sum of each row is the <u>partial</u> frequency of each number.

Example 7.1

Lotto 6/45–486 draws

A subset of 486 results from Lotto Monday draws (New South Wales—Australia) was taken at random during 2016.

The partial frequency distribution tables for this example are shown as Table 7.3A and Table 7.3B.

The sum of each column represents the number of combinations under analysis, which is 486.

The sum of each row represents the partial frequency of each number. The average partial frequency can be calculated as follows:

Because each combination has six numbers, there are

$$486 \times 6 = 2,916 \text{ numbers in the subset}$$

And because there are 45 different numbers:

Average partial frequency for every number = 2,916/45 = 64.8

The fraction of the total for the average frequency is

$$64.8/486 = 0.133$$

Example 7.2

Lotto 6/45–1,650 games

A subset of 1,650 games was made by playing Lotto Saturday (New South Wales—Australia) on different draws in 2016.

The partial frequency distribution tables for this example are shown as Table 7.4A and Table 7.4B.

The sum of each column represents the number of combinations under analysis, which is 1,650.

The sum of each row represents the partial frequency of each number. The average partial frequency can be calculated as follows:

Because each combination has six numbers, there are

$$1,650 \times 6 = 9,900 \text{ numbers in the subset}$$

And because there are 45 different numbers:

Average partial frequency for every number = 9,900/45 = 220

The fraction of the total for the average frequency is

$$220/1,650 = 0.133$$

TABLE 7.3A Partial frequency distribution table
for lotto 6/45—486 draws.

	A	B	C	D	E	F	
1	63						63
2	54	8					62
3	46	6	2				54
4	51	18	3	0			72
5	38	24	3	0	0		65
6	33	26	4	1	0	0	64
7	30	23	6	1	0	0	60
8	26	19	8	0	0	0	53
9	31	29	10	0	0	0	70
10	12	29	15	7	1	0	64
11	17	31	17	3	1	0	69
12	14	19	15	4	1	0	53
13	14	40	18	9	1	0	82
14	11	24	22	11	0	0	68
15	6	28	20	10	1	0	65
16	6	15	26	5	1	0	53
17	4	30	15	12	2	0	63
18	9	15	34	15	2	1	76
19	4	15	21	12	5	0	57
20	6	13	28	24	8	1	80
21	3	14	31	19	8	0	75
22	2	9	15	27	10	3	66
23	1	12	23	23	5	3	67
24	3	12	22	24	16	1	78
25	1	2	17	19	11	6	56
26	0	6	18	24	10	5	63
27	0	3	10	30	21	7	71
28	1	6	9	27	17	4	64
29	0	4	14	14	27	4	63
30	0	2	10	20	23	7	62
31	0	1	12	28	20	9	70
32	0	2	11	19	30	14	76
33	0	1	6	15	21	9	52
34	0	0	5	16	31	19	71
35	0	0	5	15	31	14	65
36	0	0	5	11	22	17	55
37	0	0	4	11	30	16	61
38	0	0	2	14	24	32	72
39	0	0	0	4	29	28	61
40	0	0	0	7	22	30	59
41		0	0	4	18	41	63
42			0	1	19	44	64
43				0	13	47	60
44					5	57	62
45						67	67
	486	486	486	486	486	486	2,916	64.800

TABLE 7.3B Partial frequency distribution table
for lotto 6/45 — 486 draws, with frequencies expressed
as fractions of the total.

	A	B	C	D	E	F		
1	0.130						0.130	
2	0.111	0.016					0.128	
3	0.095	0.012	0.004				0.111	
4	0.105	0.037	0.006	0.000			0.148	
5	0.078	0.049	0.006	0.000	0.000		0.134	
6	0.068	0.053	0.008	0.002	0.000	0.000	0.132	
7	0.062	0.047	0.012	0.002	0.000	0.000	0.123	
8	0.053	0.039	0.016	0.000	0.000	0.000	0.109	
9	0.064	0.060	0.021	0.000	0.000	0.000	0.144	
10	0.025	0.060	0.031	0.014	0.002	0.000	0.132	
11	0.035	0.064	0.035	0.006	0.002	0.000	0.142	
12	0.029	0.039	0.031	0.008	0.002	0.000	0.109	
13	0.029	0.082	0.037	0.019	0.002	0.000	0.169	
14	0.023	0.049	0.045	0.023	0.000	0.000	0.140	
15	0.012	0.058	0.041	0.021	0.002	0.000	0.134	
16	0.012	0.031	0.053	0.010	0.002	0.000	0.109	
17	0.008	0.062	0.031	0.025	0.004	0.000	0.130	
18	0.019	0.031	0.070	0.031	0.004	0.002	0.156	
19	0.008	0.031	0.043	0.025	0.010	0.000	0.117	
20	0.012	0.027	0.058	0.049	0.016	0.002	0.165	
21	0.006	0.029	0.064	0.039	0.016	0.000	0.154	
22	0.004	0.019	0.031	0.056	0.021	0.006	0.136	
23	0.002	0.025	0.047	0.047	0.010	0.006	0.138	
24	0.006	0.025	0.045	0.049	0.033	0.002	0.160	
25	0.002	0.004	0.035	0.039	0.023	0.012	0.115	
26	0.000	0.012	0.037	0.049	0.021	0.010	0.130	
27	0.000	0.006	0.021	0.062	0.043	0.014	0.146	
28	0.002	0.012	0.019	0.056	0.035	0.008	0.132	
29	0.000	0.008	0.029	0.029	0.056	0.008	0.130	
30	0.000	0.004	0.021	0.041	0.047	0.014	0.128	
31	0.000	0.002	0.025	0.058	0.041	0.019	0.144	
32	0.000	0.004	0.023	0.039	0.062	0.029	0.156	
33	0.000	0.002	0.012	0.031	0.043	0.019	0.107	
34	0.000	0.000	0.010	0.033	0.064	0.039	0.146	
35	0.000	0.000	0.010	0.031	0.064	0.029	0.134	
36	0.000	0.000	0.010	0.023	0.045	0.035	0.113	
37	0.000	0.000	0.008	0.023	0.062	0.033	0.126	
38	0.000	0.000	0.004	0.029	0.049	0.066	0.148	
39	0.000	0.000	0.000	0.008	0.060	0.058	0.126	
40	0.000	0.000	0.000	0.014	0.045	0.062	0.121	
41		0.000	0.000	0.008	0.037	0.084	0.130	
42			0.000	0.002	0.039	0.091	0.132	
43				0.000	0.027	0.097	0.123	
44					0.010	0.117	0.128	
45						0.138	0.138	
	1.000	1.000	1.000	1.000	1.000	1.000	6.000	0.133

TABLE 7.4A Partial frequency distribution table
for lotto 6/45 — 1,650 games.

	A	B	C	D	E	F		
1	234						234	
2	191	22					213	
3	164	45	2				211	
4	145	65	8	0			218	
5	113	81	10	0	0		204	
6	126	79	19	3	0	0	227	
7	100	98	18	5	0	0	221	
8	93	84	39	3	0	0	219	
9	83	73	40	9	0	0	205	
10	78	107	40	5	1	0	231	
11	· 49	99	54	16	1	0	219	
12	53	87	51	17	3	0	211	
13	38	99	71	25	6	0	239	
14	37	79	69	29	3	1	218	
15	18	76	83	27	5	1	210	
16	25	73	76	44	11	1	230	
17	19	68	85	35	5	1	213	
18	21	64	82	62	15	1	245	
19	16	52	76	44	18	0	206	
20	10	50	89	51	15	3	218	
21	7	36	87	72	28	5	235	
22	6	28	78	82	24	2	220	
23	3	47	60	65	30	4	209	
24	4	27	69	71	36	5	212	
25	5	19	61	63	43	6	197	
26	2	20	70	85	41	10	228	
27	4	17	47	93	40	22	223	
28	1	12	43	81	66	11	214	
29	1	14	49	79	76	11	230	
30	2	6	40	75	74	33	230	
31	0	9	25	72	74	24	204	
32	0	3	24	79	77	28	211	
33	0	3	24	65	90	43	225	
34	1	1	19	52	104	38	215	
35	0	3	12	52	99	68	234	
36	1	2	8	40	101	76	228	
37	0	1	3	35	93	77	209	
38	0	1	6	34	102	97	240	
39	0	0	5	25	75	95	200	
40	0	0	7	19	80	124	230	
41		0	0	20	72	128	220	
42			1	13	74	135	223	
43				3	48	167	218	
44					20	210	230	
45						223	223	
	1,650	1,650	1,650	1,650	1,650	1,650	9,900	220.000

TABLE 7.4B Partial frequency distribution table
for lotto 6/45—1,650 games, with frequencies expressed
as fractions of the total.

	A	B	C	D	E	F		
1	0.142						0.142	
2	0.116	0.013					0.129	
3	0.099	0.027	0.001				0.128	
4	0.088	0.039	0.005	0.000			0.132	
5	0.068	0.049	0.006	0.000	0.000		0.124	
6	0.076	0.048	0.012	0.002	0.000	0.000	0.138	
7	0.061	0.059	0.011	0.003	0.000	0.000	0.134	
8	0.056	0.051	0.024	0.002	0.000	0.000	0.133	
9	0.050	0.044	0.024	0.005	0.000	0.000	0.124	
10	0.047	0.065	0.024	0.003	0.001	0.000	0.140	
11	0.030	0.060	0.033	0.010	0.001	0.000	0.133	
12	0.032	0.053	0.031	0.010	0.002	0.000	0.128	
13	0.023	0.060	0.043	0.015	0.004	0.000	0.145	
14	0.022	0.048	0.042	0.018	0.002	0.001	0.132	
15	0.011	0.046	0.050	0.016	0.003	0.001	0.127	
16	0.015	0.044	0.046	0.027	0.007	0.001	0.139	
17	0.012	0.041	0.052	0.021	0.003	0.001	0.129	
18	0.013	0.039	0.050	0.038	0.009	0.001	0.148	
19	0.010	0.032	0.046	0.027	0.011	0.000	0.125	
20	0.006	0.030	0.054	0.031	0.009	0.002	0.132	
21	0.004	0.022	0.053	0.044	0.017	0.003	0.142	
22	0.004	0.017	0.047	0.050	0.015	0.001	0.133	
23	0.002	0.028	0.036	0.039	0.018	0.002	0.127	
24	0.002	0.016	0.042	0.043	0.022	0.003	0.128	
25	0.003	0.012	0.037	0.038	0.026	0.004	0.119	
26	0.001	0.012	0.042	0.052	0.025	0.006	0.138	
27	0.002	0.010	0.028	0.056	0.024	0.013	0.135	
28	0.001	0.007	0.026	0.049	0.040	0.007	0.130	
29	0.001	0.008	0.030	0.048	0.046	0.007	0.139	
30	0.001	0.004	0.024	0.045	0.045	0.020	0.139	
31	0.000	0.005	0.015	0.044	0.045	0.015	0.124	
32	0.000	0.002	0.015	0.048	0.047	0.017	0.128	
33	0.000	0.002	0.015	0.039	0.055	0.026	0.136	
34	0.001	0.001	0.012	0.032	0.063	0.023	0.130	
35	0.000	0.002	0.007	0.032	0.060	0.041	0.142	
36	0.001	0.001	0.005	0.024	0.061	0.046	0.138	
37	0.000	0.001	0.002	0.021	0.056	0.047	0.127	
38	0.000	0.001	0.004	0.021	0.062	0.059	0.145	
39	0.000	0.000	0.003	0.015	0.045	0.058	0.121	
40	0.000	0.000	0.004	0.012	0.048	0.075	0.139	
41		0.000	0.000	0.012	0.044	0.078	0.133	
42			0.001	0.008	0.045	0.082	0.135	
43				0.002	0.029	0.101	0.132	
44					0.012	0.127	0.139	
45						0.135	0.135	
	1.000	1.000	1.000	1.000	1.000	1.000	**6.000**	**0.133**

Example 7.3

<u>Lotto 6/45—1,508 draws</u>

A subset of 1,508 results from Megalotto 6/45 draws (Philippines) was taken at random during 2016.

The partial frequency distribution tables for this example are shown as Table 7.5A and Table 7.5B.

The sum of each column represents the number of combinations under analysis, which is 1,508.

The sum of each row represents the partial frequency of each number. The average partial frequency can be calculated as follows:

Because each combination has six numbers, there are

$$1,508 \times 6 = 9,048 \text{ numbers in the subset}$$

And because there are 45 different numbers:

Average partial frequency for every number = 9,048/45 = 201.067

The fraction of the total for the average frequency is

$$201.067/1,508 = 0.133$$

TABLE 7.5A Partial frequency distribution table for lotto 6/45 — 1,508 draws.

	A	B	C	D	E	F	
1	195						195
2	160	23					183
3	175	38	2				215
4	134	66	5	0			205
5	110	63	10	0	0		183
6	98	88	18	3	0	0	207
7	85	70	28	1	0	0	184
8	93	69	28	3	0	0	193
9	65	74	32	7	1	0	179
10	65	87	47	9	0	0	208
11	52	103	46	8	1	0	210
12	59	85	47	18	2	1	212
13	45	81	58	24	4	0	212
14	34	77	62	28	3	0	204
15	34	83	79	29	5	1	231
16	21	64	72	29	8	1	195
17	11	72	72	43	13	1	212
18	5	53	76	47	9	0	190
19	10	42	89	42	10	1	194
20	14	37	80	47	16	1	195
21	8	36	64	62	32	3	205
22	11	41	53	54	23	1	183
23	9	27	67	78	29	3	213
24	4	21	57	66	34	6	188
25	5	18	74	63	38	8	206
26	2	33	47	72	50	12	216
27	1	15	42	75	48	7	188
28	0	11	51	76	48	11	197
29	1	7	50	74	70	27	229
30	0	6	26	85	77	29	223
31	2	8	28	64	55	29	186
32	0	4	27	72	77	37	217
33	0	1	21	58	87	37	204
34	0	4	15	59	89	42	209
35	0	1	8	44	82	54	189
36	0	0	11	41	77	58	187
37	0	0	5	35	93	68	201
38	0	0	6	33	91	82	212
39	0	0	5	20	82	86	193
40	0	0	0	19	72	95	186
41		0	0	15	52	106	173
42			0	5	65	133	203
43				0	38	174	212
44					27	179	206
45						215	215
	1,508	1,508	1,508	1,508	1,508	1,508	9,048	201.067

TABLE 7.5B Partial frequency distribution table
for lotto 6/45—1,508 draws, with frequencies expressed
as fractions of the total.

	A	B	C	D	E	F		
1	0.129						0.129	
2	0.106	0.015					0.121	
3	0.116	0.025	0.001				0.143	
4	0.089	0.044	0.003	0.000			0.136	
5	0.073	0.042	0.007	0.000	0.000		0.121	
6	0.065	0.058	0.012	0.002	0.000	0.000	0.137	
7	0.056	0.046	0.019	0.001	0.000	0.000	0.122	
8	0.062	0.046	0.019	0.002	0.000	0.000	0.128	
9	0.043	0.049	0.021	0.005	0.001	0.000	0.119	
10	0.043	0.058	0.031	0.006	0.000	0.000	0.138	
11	0.034	0.068	0.031	0.005	0.001	0.000	0.139	
12	0.039	0.056	0.031	0.012	0.001	0.001	0.141	
13	0.030	0.054	0.038	0.016	0.003	0.000	0.141	
14	0.023	0.051	0.041	0.019	0.002	0.000	0.135	
15	0.023	0.055	0.052	0.019	0.003	0.001	0.153	
16	0.014	0.042	0.048	0.019	0.005	0.001	0.129	
17	0.007	0.048	0.048	0.029	0.009	0.001	0.141	
18	0.003	0.035	0.050	0.031	0.006	0.000	0.126	
19	0.007	0.028	0.059	0.028	0.007	0.001	0.129	
20	0.009	0.025	0.053	0.031	0.011	0.001	0.129	
21	0.005	0.024	0.042	0.041	0.021	0.002	0.136	
22	0.007	0.027	0.035	0.036	0.015	0.001	0.121	
23	0.006	0.018	0.044	0.052	0.019	0.002	0.141	
24	0.003	0.014	0.038	0.044	0.023	0.004	0.125	
25	0.003	0.012	0.049	0.042	0.025	0.005	0.137	
26	0.001	0.022	0.031	0.048	0.033	0.008	0.143	
27	0.001	0.010	0.028	0.050	0.032	0.005	0.125	
28	0.000	0.007	0.034	0.050	0.032	0.007	0.131	
29	0.001	0.005	0.033	0.049	0.046	0.018	0.152	
30	0.000	0.004	0.017	0.056	0.051	0.019	0.148	
31	0.001	0.005	0.019	0.042	0.036	0.019	0.123	
32	0.000	0.003	0.018	0.048	0.051	0.025	0.144	
33	0.000	0.001	0.014	0.038	0.058	0.025	0.135	
34	0.000	0.003	0.010	0.039	0.059	0.028	0.139	
35	0.000	0.001	0.005	0.029	0.054	0.036	0.125	
36	0.000	0.000	0.007	0.027	0.051	0.038	0.124	
37	0.000	0.000	0.003	0.023	0.062	0.045	0.133	
38	0.000	0.000	0.004	0.022	0.060	0.054	0.141	
39	0.000	0.000	0.003	0.013	0.054	0.057	0.128	
40	0.000	0.000	0.000	0.013	0.048	0.063	0.123	
41		0.000	0.000	0.010	0.034	0.070	0.115	
42			0.000	0.003	0.043	0.088	0.135	
43				0.000	0.025	0.115	0.141	
44					0.018	0.119	0.137	
45						0.143	0.143	
	1.000	1.000	1.000	1.000	1.000	1.000	6.000	0.133

Example 7.4

Lotto 6/45—8,145 draws

Given that one type of lottery is the same no matter when and where it is played, a subset of 8,145 draws for 6/45 was made by grouping all results of subsets taken from the following lotteries:

Lottery	Number of draws
Megalotto 6/45 (Philippines)	1,512
Wednesday Gold Lotto (Queensland—Australia)	1,236
Baloto 6/45 (Colombia)	1,088
Croatia Lotto 6/45 (Croatia)	658
Oz Lotto 6/45 (Australia)	608
Saturday Lotto (New South Wales—Australia)	563
Irish Lotto (Ireland)	511
Lotto Plus 1 (Ireland)	494
Lotto Plus 2 (Ireland)	494
Monday Lotto (New South Wales—Australia)	388
Wednesday Lotto (New South Wales—Australia)	388
Ontario Lottario (Canada)	205
Total:	**8,145**

NOTE: all of the aforementioned lotteries followed a 6/45 format at the time when the data was collected (randomly during 2016).

A subset of 8,145 draws was taken as shown in the table above.

The partial frequency distribution tables for this example are shown as Table 7.5A and Table 7.5B.

The sum of each column represents the number of combinations under analysis, which is 8,145.

The sum of each row represents the partial frequency of each number. The average partial frequency can be calculated as follows:

Because each combination has six numbers, there are

$$8{,}145 \times 6 = 48{,}870 \text{ numbers in the subset}$$

And because there are 45 different numbers:

Average partial frequency for every number = 48,870/45 = 1,086

The fraction of the total for the average frequency is

$$1{,}086/8{,}145 = 0.133$$

TABLE 7.6A Partial frequency distribution table for lotto 6/45 — 8,145 draws.

	A	B	C	D	E	F		
1	1,071						1,071	
2	977	130					1,107	
3	921	231	8				1,160	
4	746	318	31	1			1,096	
5	621	379	60	4	0		1,064	
6	571	442	93	6	0	0	1,112	
7	518	428	141	13	0	0	1,100	
8	455	404	149	25	0	0	1,033	
9	378	441	187	29	3	0	1,038	
10	295	530	247	61	1	0	1,134	
11	286	497	278	69	6	0	1,136	
12	228	466	305	82	8	1	1,090	
13	190	404	333	117	17	1	1,062	
14	168	378	342	141	21	0	1,050	
15	174	406	387	152	29	3	1,151	
16	116	336	393	189	39	1	1,074	
17	84	360	382	210	53	7	1,096	
18	69	275	396	264	63	5	1,072	
19	60	242	412	244	86	4	1,048	
20	42	242	367	298	97	17	1,063	
21	42	231	399	313	120	20	1,125	
22	26	191	330	347	143	17	1,054	
23	25	163	367	364	159	27	1,105	
24	18	127	341	389	175	31	1,081	
25	16	105	324	346	211	51	1,053	
26	12	101	310	383	236	57	1,099	
27	14	78	254	420	302	70	1,138	
28	8	59	236	380	307	72	1,062	
29	4	46	231	381	347	105	1,114	
30	1	36	153	382	382	120	1,074	
31	4	28	169	357	358	148	1,064	
32	2	28	127	324	392	175	1,048	
33	0	18	107	348	415	189	1,077	
34	3	12	75	317	422	226	1,055	
35	0	6	70	271	492	275	1,114	
36	0	2	53	217	454	306	1,032	
37	0	3	34	193	460	376	1,066	
38	0	1	31	196	480	444	1,152	
39	0	1	19	101	466	514	1,101	
40	0	0	3	120	403	594	1,120	
41		0	0	58	335	648	1,041	
42			1	26	327	714	1,068	
43				7	206	866	1,079	
44					130	926	1,056	
45						1,135	1,135	
	8,145	8,145	8,145	8,145	8,145	8,145	48,870	1,086.000

TABLE 7.6B Partial frequency distribution table
for lotto 6/45 — 8,145 draws, with frequencies expressed
as fractions of the total.

	A	B	C	D	E	F	
1	0.131						0.131
2	0.120	0.016					0.136
3	0.113	0.028	0.001				0.142
4	0.092	0.039	0.004	0.000			0.135
5	0.076	0.047	0.007	0.000	0.000		0.131
6	0.070	0.054	0.011	0.001	0.000	0.000	0.137
7	0.064	0.053	0.017	0.002	0.000	0.000	0.135
8	0.056	0.050	0.018	0.003	0.000	0.000	0.127
9	0.046	0.054	0.023	0.004	0.000	0.000	0.127
10	0.036	0.065	0.030	0.007	0.000	0.000	0.139
11	0.035	0.061	0.034	0.008	0.001	0.000	0.139
12	0.028	0.057	0.037	0.010	0.001	0.000	0.134
13	0.023	0.050	0.041	0.014	0.002	0.000	0.130
14	0.021	0.046	0.042	0.017	0.003	0.000	0.129
15	0.021	0.050	0.048	0.019	0.004	0.000	0.141
16	0.014	0.041	0.048	0.023	0.005	0.000	0.132
17	0.010	0.044	0.047	0.026	0.007	0.001	0.135
18	0.008	0.034	0.049	0.032	0.008	0.001	0.132
19	0.007	0.030	0.051	0.030	0.011	0.000	0.129
20	0.005	0.030	0.045	0.037	0.012	0.002	0.131
21	0.005	0.028	0.049	0.038	0.015	0.002	0.138
22	0.003	0.023	0.041	0.043	0.018	0.002	0.129
23	0.003	0.020	0.045	0.045	0.020	0.003	0.136
24	0.002	0.016	0.042	0.048	0.021	0.004	0.133
25	0.002	0.013	0.040	0.042	0.026	0.006	0.129
26	0.001	0.012	0.038	0.047	0.029	0.007	0.135
27	0.002	0.010	0.031	0.052	0.037	0.009	0.140
28	0.001	0.007	0.029	0.047	0.038	0.009	0.130
29	0.000	0.006	0.028	0.047	0.043	0.013	0.137
30	0.000	0.004	0.019	0.047	0.047	0.015	0.132
31	0.000	0.003	0.021	0.044	0.044	0.018	0.131
32	0.000	0.003	0.016	0.040	0.048	0.021	0.129
33	0.000	0.002	0.013	0.043	0.051	0.023	0.132
34	0.000	0.001	0.009	0.039	0.052	0.028	0.130
35	0.000	0.001	0.009	0.033	0.060	0.034	0.137
36	0.000	0.000	0.007	0.027	0.056	0.038	0.127
37	0.000	0.000	0.004	0.024	0.056	0.046	0.131
38	0.000	0.000	0.004	0.024	0.059	0.055	0.141
39	0.000	0.000	0.002	0.012	0.057	0.063	0.135
40	0.000	0.000	0.000	0.015	0.049	0.073	0.138
41		0.000	0.000	0.007	0.041	0.080	0.128
42		0.000	0.000	0.003	0.040	0.088	0.131
43			0.001	0.025	0.106		0.132
44				0.016	0.114		0.130
45					0.139		0.139
	1.000	1.000	1.000	1.000	1.000	1.000	6.000	0.133

Example 7.5

Lotto 5/39—5,757 draws

A subset of 5,757 results from Fantasy 5 draws (Georgia, USA) was taken at random during 2016.

The partial frequency distribution tables for this example are shown as Table 7.6A and Table 7.6B.

The sum of each column represents the number of combinations under analysis, which is 5,757.

The sum of each row represents the partial frequency of each number. The average partial frequency can be calculated as follows:

Because each combination has six numbers, there are

$$5{,}757 \times 5 = 28{,}785 \text{ numbers in the subset}$$

And because there are 39 different numbers:

$$\text{Frequency for every number} = 28{,}785/39 = 738.077$$

The fraction of the total for the average frequency is

$$738.077/5{,}757 = 0.128$$

TABLE 7.7A Partial frequency distribution table
for lotto 5/39 — 5,757 draws.

	A	B	C	D	E		
1	782						782
2	693	87					780
3	620	145	7				772
4	517	199	30	0			746
5	474	233	29	1	0		737
6	401	305	68	2	0		776
7	378	313	92	13	1		797
8	297	346	131	18	1		793
9	266	314	138	19	1		738
10	239	355	137	28	2		761
11	199	313	202	30	1		745
12	148	324	190	66	5		733
13	132	331	216	57	9		745
14	122	274	239	77	7		719
15	98	299	258	93	11		759
16	79	265	292	112	20		768
17	62	249	261	144	26		742
18	58	221	321	182	37		819
19	51	199	319	185	34		788
20	36	206	320	194	46		802
21	29	142	290	253	59		773
22	21	108	280	257	71		737
23	14	111	277	248	93		743
24	10	103	230	314	110		767
25	16	87	231	287	135		756
26	8	72	231	306	147		764
27	2	53	223	335	175		788
28	4	35	169	377	209		794
29	0	29	150	326	276		781
30	1	12	137	306	293		749
31	0	12	95	337	341		785
32	0	8	75	243	359		685
33	0	4	48	288	467		807
34	0	3	30	211	490		734
35	0	0	22	161	594		777
36		0	13	125	355		493
37			6	115	437		558
38				47	458		505
39					487		487
	5,757	5,757	5,757	5,757	5,757	28,785	738.077

TABLE 7.7B Partial frequency distribution table
for lotto 5/39—5,757 draws, with frequencies expressed
as fractions of the total.

	A	B	C	D	E		
1	0.136					0.136	
2	0.120	0.015				0.135	
3	0.108	0.025	0.001			0.134	
4	0.090	0.035	0.005	0.000		0.130	
5	0.082	0.040	0.005	0.000	0.000	0.128	
6	0.070	0.053	0.012	0.000	0.000	0.135	
7	0.066	0.054	0.016	0.002	0.000	0.138	
8	0.052	0.060	0.023	0.003	0.000	0.138	
9	0.046	0.055	0.024	0.003	0.000	0.128	
10	0.042	0.062	0.024	0.005	0.000	0.132	
11	0.035	0.054	0.035	0.005	0.000	0.129	
12	0.026	0.056	0.033	0.011	0.001	0.127	
13	0.023	0.057	0.038	0.010	0.002	0.129	
14	0.021	0.048	0.042	0.013	0.001	0.125	
15	0.017	0.052	0.045	0.016	0.002	0.132	
16	0.014	0.046	0.051	0.019	0.003	0.133	
17	0.011	0.043	0.045	0.025	0.005	0.129	
18	0.010	0.038	0.056	0.032	0.006	0.142	
19	0.009	0.035	0.055	0.032	0.006	0.137	
20	0.006	0.036	0.056	0.034	0.008	0.139	
21	0.005	0.025	0.050	0.044	0.010	0.134	
22	0.004	0.019	0.049	0.045	0.012	0.128	
23	0.002	0.019	0.048	0.043	0.016	0.129	
24	0.002	0.018	0.040	0.055	0.019	0.133	
25	0.003	0.015	0.040	0.050	0.023	0.131	
26	0.001	0.013	0.040	0.053	0.026	0.133	
27	0.000	0.009	0.039	0.058	0.030	0.137	
28	0.001	0.006	0.029	0.065	0.036	0.138	
29	0.000	0.005	0.026	0.057	0.048	0.136	
30	0.000	0.002	0.024	0.053	0.051	0.130	
31	0.000	0.002	0.017	0.059	0.059	0.136	
32	0.000	0.001	0.013	0.042	0.062	0.119	
33	0.000	0.001	0.008	0.050	0.081	0.140	
34	0.000	0.001	0.005	0.037	0.085	0.127	
35	0.000	0.000	0.004	0.028	0.103	0.135	
36		0.000	0.002	0.022	0.062	0.086	
37			0.001	0.020	0.076	0.097	
38				0.008	0.080	0.088	
39					0.085	0.085	
	1.000	1.000	1.000	1.000	1.000	5.000	0.128

Example 7.6

Lotto 5/39—7,023 draws

A subset of 7,023 results from Fantasy 5 (California) draws was taken at random during 2016.

The partial frequency distribution tables for this example are shown as Table 7.7A and Table 7.7B.

The sum of each column represents the number of combinations under analysis, which is 7,023.

The sum of each row represents the partial frequency of each number. The average frequency can be calculated as follows:

Because each combination has five numbers, there are

$$7,023 \times 5 = 35,115 \text{ numbers in the whole subset}$$

And because there are 39 different numbers:

Average partial frequency for every number = 35,515/39 = 900.38

The fraction of the total for the average frequency is

$$900.38/7,023 = 0.128$$

TABLE 7.8A Partial frequency distribution table
for lotto 5/39—7,023 draws.

	A	B	C	D	E		
1	885						885
2	842	112					954
3	730	185	11				926
4	604	251	24	1			880
5	592	294	40	1	0		927
6	510	342	71	3	0		926
7	426	377	100	15	0		918
8	358	393	112	14	0		877
9	334	391	156	23	2		906
10	275	417	200	32	1		925
11	265	389	216	46	3		919
12	215	389	228	58	8		898
13	197	399	280	73	8		957
14	153	354	297	77	5		886
15	125	364	279	108	13		889
16	88	293	340	128	22		871
17	75	271	318	126	30		820
18	81	237	345	183	29		875
19	64	233	327	196	31		851
20	52	240	339	215	45		891
21	37	177	337	242	76		869
22	30	202	313	267	75		887
23	25	130	346	308	85		894
24	20	131	349	335	107		942
25	11	123	315	353	139		941
26	14	94	288	354	136		886
27	8	74	286	366	160		894
28	5	44	245	390	228		912
29	1	32	213	403	236		885
30	1	34	183	411	303		932
31	0	19	155	388	353		915
32	0	17	113	405	391		926
33	0	12	77	369	418		876
34	0	3	70	365	486		924
35	0	0	27	301	564		892
36		0	13	221	664		898
37			10	159	733		902
38				87	748		835
39					924		924
	7,023	7,023	7,023	7,023	7,023	35,115	900.385

TABLE 7.8B Partial frequency distribution table for lotto 5/39—7,023 draws, with frequencies expressed as fractions of the total.

	A	B	C	D	E		
1	0.126					0.126	
2	0.120	0.016				0.136	
3	0.104	0.026	0.002			0.132	
4	0.086	0.036	0.003	0.000		0.125	
5	0.084	0.042	0.006	0.000	0.000	0.132	
6	0.073	0.049	0.010	0.000	0.000	0.132	
7	0.061	0.054	0.014	0.002	0.000	0.131	
8	0.051	0.056	0.016	0.002	0.000	0.125	
9	0.048	0.056	0.022	0.003	0.000	0.129	
10	0.039	0.059	0.028	0.005	0.000	0.132	
11	0.038	0.055	0.031	0.007	0.000	0.131	
12	0.031	0.055	0.032	0.008	0.001	0.128	
13	0.028	0.057	0.040	0.010	0.001	0.136	
14	0.022	0.050	0.042	0.011	0.001	0.126	
15	0.018	0.052	0.040	0.015	0.002	0.127	
16	0.013	0.042	0.048	0.018	0.003	0.124	
17	0.011	0.039	0.045	0.018	0.004	0.117	
18	0.012	0.034	0.049	0.026	0.004	0.125	
19	0.009	0.033	0.047	0.028	0.004	0.121	
20	0.007	0.034	0.048	0.031	0.006	0.127	
21	0.005	0.025	0.048	0.034	0.011	0.124	
22	0.004	0.029	0.045	0.038	0.011	0.126	
23	0.004	0.019	0.049	0.044	0.012	0.127	
24	0.003	0.019	0.050	0.048	0.015	0.134	
25	0.002	0.018	0.045	0.050	0.020	0.134	
26	0.002	0.013	0.041	0.050	0.019	0.126	
27	0.001	0.011	0.041	0.052	0.023	0.127	
28	0.001	0.006	0.035	0.056	0.032	0.130	
29	0.000	0.005	0.030	0.057	0.034	0.126	
30	0.000	0.005	0.026	0.059	0.043	0.133	
31	0.000	0.003	0.022	0.055	0.050	0.130	
32	0.000	0.002	0.016	0.058	0.056	0.132	
33	0.000	0.002	0.011	0.053	0.060	0.125	
34	0.000	0.000	0.010	0.052	0.069	0.132	
35	0.000	0.000	0.004	0.043	0.080	0.127	
36		0.000	0.002	0.031	0.095	0.128	
37			0.001	0.023	0.104	0.128	
38				0.012	0.107	0.119	
39					0.132	0.132	
	1.000	1.000	1.000	1.000	1.000	5.000	0.128

Conclusion

When the frequencies for any subset (a specified number of draws) are expressed as a fraction of the total, they have a value close to the frequency of the set (total possible outcomes), when the set is also expressed as a fraction of the total. This means that the frequency of every number in any lottery subset is about the same. The larger the subset (as the number of draws increases), the closer is the frequency of each number to uniformity. The average frequency has a trend to reach the value of the absolute frequency.

Any subset (any lottery) follows the same pattern. Many other examples can be made with different lotteries and the results will always confirm this.

Subsets as small as 0.006% of the total number of possible outcomes follow the trend — see examples in table 7.9.

In conclusion, the partial frequency for each number when a fraction of all possible combinations of C(n,r) is evaluated has the tendency to reach the average theoretical frequency.

In other words, every lottery without replacement is a subset of C(n,r), and intrinsically it follows the same frequency distribution as the theoretical set. Therefore, the appearance of "most frequent numbers" is only transitory, and it must not be taken as mathematical prediction of the occurrence of particular numbers in future draws.

TABLE 7.9 Frequency distributions shown as percentages
of the total number of possible combinations

Lottery	Number of Draws/Games	Per Cent of Total Possible Combinations	Average Frequency (As a fraction of the total)
6/45	8,145,060	100	0.133
6/45	8,145	0.100	0.133
6/45	1,650	0.020	0.133
6/45	1,508	0.019	0.133
6/45	486	0.006	0.133
5/39	575,757	100	0.128
5/39	7,023	1.220	0.128
5/39	5,757	1.000	0.128

7.3 Lotto 1/350,000

Unfairness in a lottery may occur when the favorable outcomes are not well defined. The following is a real-life example in which such situation happened.

A random number generator selects a six-digit number in the range from 100,000 to 449,999. The tickets are also randomly generated, the player does not have the option to choose the numbers.

PRIZE STRUCTURE	Probability 1 in
Division 1: Match the six-digit number in the exact order drawn.	350,000
Division 2: Match the last five digits in the exact order drawn.	?
Division 3: Match the last four digits in the exact order drawn.	?
Division 4: Match the last three digits in the exact order drawn.	1,111

There is unfairness in this lottery because Division 2 and Division 3 have different probabilities, depending on whether the player is allocated a number where the second digit is 0 to 4, or a number where the second digit is 0 to 9. The player does not have the option to choose, given the fact that all tickets are randomly generated.

This type of lottery (See figure 7.1) was played in New South Wales (Australia) from 11 October 1996 to 8 December 2000 on a weekly basis, with a total of 218 draws. There were 126 draws with either of the digits 0 to 4 as the second number in the winning numbers, and 92 draws with either of the digits 5 to 9 as the second number in the winning numbers. The digit 4 was drawn 30 times as the first digit in the winning numbers.

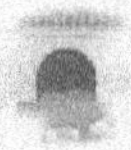
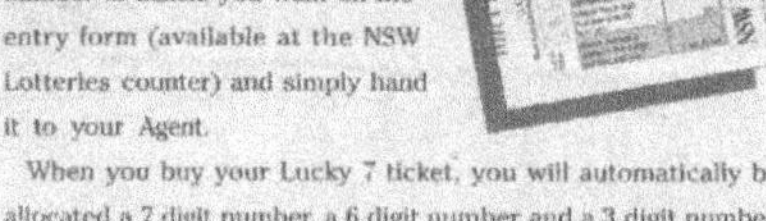

What is Lucky 7?

Lucky 7 is the new Lucky Lotteries product which gives you five chances to win $1 million! The ticket costs $10 (including Agent's commission) and lasts for five weekly draws. Lucky 7 gives you seven ways to win each week.

How to Play

If you just want to buy one Lucky 7 ticket, all you have to do is ask your nearest on-line NSW Lotteries Agent for a ticket.

If you want to buy more than one Lucky 7 ticket, just fill in the number of tickets you want on the entry form (available at the NSW Lotteries counter) and simply hand it to your Agent.

When you buy your Lucky 7 ticket, you will automatically be allocated a 7 digit number, a 6 digit number and a 3 digit number, which will be valid for the next five draws. The first Lucky 7 draw shown on your ticket will be held the Friday following your purchase. The remaining draws will be held each Friday over the next four weeks.

How is it Drawn?

Unlike the $2 and $5 Jackpot Lotteries, all the tickets do not have to be sold in Lucky 7 for the lottery to be drawn.

In this respect, Lucky 7 works more like a Lotto style game, but you don't have to share your prize! The same random number generator that is currently used every weekday at NSW Lotteries Head Office in Burwood will select the Lucky 7 winning numbers each Friday. Any member of the public can draw Lucky 7.

How to Win

Lucky 7 gives you seven ways to win in each of the 5 draws for which your ticket is valid.

Lucky 7:
1 4 5 2 3 6 9
Match the 7 digit number exactly and WIN $1,000,000

Lucky 6:
3 5 2 4 7 1
Match the 6 digit number exactly and WIN $20,000
4 5 2 4 7 1
Match the last 5 digits of the 6 digit number and WIN $2,000
2 1 2 4 7 1
Match the last 4 digits of the 6 digit number and WIN $200
2 3 5 4 7 1
Match the last 3 digits of the 6 digit number and WIN $20

Lucky 3:
4 5 9
Match the 3 digit number exactly and WIN $20
1 5 9
Match the last 2 digits of the 3 digit number and WIN A FREE LUCKY 7 TICKET

Note: If you match the 6 digit number exactly you cannot claim the lower prize levels relating to the match of the last 5 digits and the last 4 digits of the 6 digit number.

If you match the 3 digit number exactly you cannot claim the lower prize levels relating to the match of the last 2 digits of the 3 digit number.

Facts About Lucky 7

Your Lucky 7 number will be allocated in the range from 1,000,000 to 1,999,999. Your Lucky 6 number will be allocated in the range from 100,000 to 449,999. Your Lucky 3 number will be allocated in the range from 1 to 999.

The odds of winning any prize is 1 in 18. The odds of winning Lucky 7 are 1 in 200,000. The odds of winning Lucky 6 are 1 in 70,000. The odds of winning Lucky 3 are 1 in 200.

FIGURE 7.1 Lotto 1/350,000 (Lucky 6) as a sub-division of Lucky 7
(Reproduced with permission of Tatts Group—Australia)

Facts About Lucky 7

Your Lucky 7 number will be allocated in the range from 1,000,000 to 1,999,999. Your Lucky 6 number will be allocated in the range from 100,000 to 449,999. Your Lucky 3 number will be allocated in the range from 1 to 999.

The odds of winning any prize is 1 in 18. The odds of winning Lucky 7 are 1 in 200,000. The odds of winning Lucky 6 are 1 in 70,000. The odds of winning Lucky 3 are 1 in 200.

100,000 to 449,999

1 in 70,000

The Division 1 probability is 1 in 70,000 because one ticket was valid for 5 draws, hence $P(\text{Div. 1}) = 1 \text{ in } \frac{350,000}{5} = 1 \text{ in } 70,000$

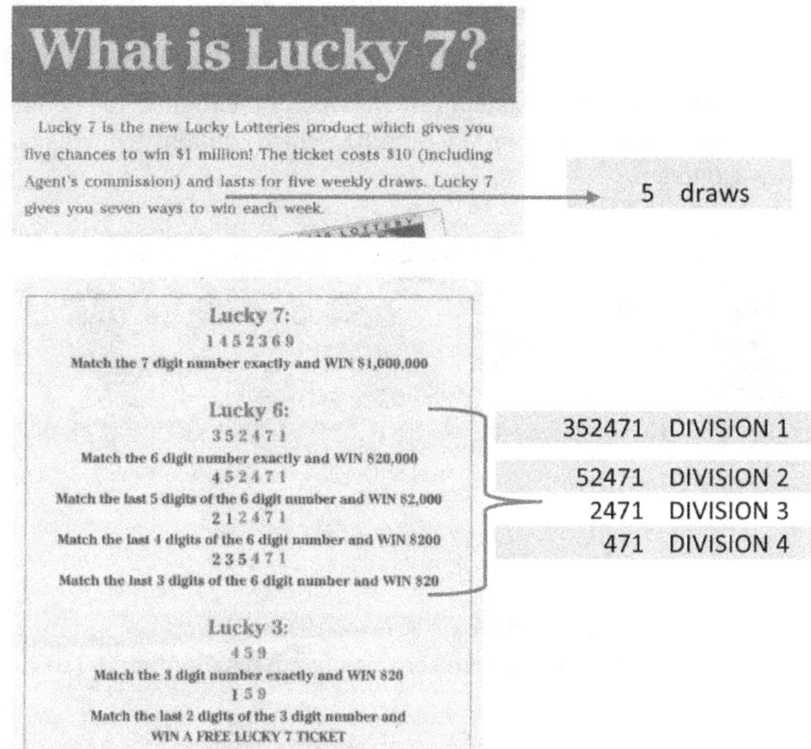

The above is an example of Lucky 6 for the winning number 352741.

Once the Division 1 prize is drawn, a difficulty arises to work out he probabilities for Divisions 2 and 3.

Let us visualise this lottery as if the draws were performed by using lottery ball drawing machines:

Six balls are drawn, one from each of six lottery ball drawing machines, each containing a defined number of balls, as follows:

Machine **A** contains <u>four</u> balls numbered from 1 to 4

Machine **B1** contains <u>ten</u> balls numbered from 0 to 9.

Machine **B2** contains <u>five</u> balls numbered from 0 to 4.

Machine **C** contains <u>ten</u> balls numbered from 0 to 9

Machine **D** contains <u>ten</u> balls numbered from 0 to 9

Machine **E** contains <u>ten</u> balls numbered from 0 to 9

Machine **F** contains <u>ten</u> balls numbered from 0 to 9

The draw of the second number is dependent on the number drawn from Machine A, as follows:

SITUATION A

If the number drawn from Machine **A** is either of the digits 1 to 3, then a second number is drawn from Machine **B1**, which contains <u>ten</u> balls numbered 0 to 9.

PROGRESSION 1

A	B1	C	D	E	F
1 2 3	0 1 2 3 4 5 6 7 8 9	0 1 2 3 4 5 6 7 8 9	0 1 2 3 4 5 6 7 8 9	0 1 2 3 4 5 6 7 8 9	0 1 2 3 4 5 6 7 8 9

SITUATION B

If the number drawn from Machine **A** is the digit 4, then a second number is drawn from Machine **B2**, which contains <u>five</u> balls numbered 0 to 4.

PROGRESSION 2

A	B2	C	D	E	F
4	0 1 2 3 4	0 1 2 3 4 5 6 7 8 9	0 1 2 3 4 5 6 7 8 9	0 1 2 3 4 5 6 7 8 9	0 1 2 3 4 5 6 7 8 9

There is an evident problem with the second digit: if the first number drawn is either of the digits 1 to 3, and the second digit printed on your ticket is either of the digits 0 to 9 (Situation A), you are ipso facto missing 50% of your chances to match the last five digits!

To better understand this fact and work out the probabilities of matching the last five, four and three digits on Lucky 6, each situation must be considered as a separate lottery.

Once the six-digit number is drawn the player is allocated a number where the second digit is 0 to 9, or a number where the second digit is 0 to 4 (the player does not have the option to choose, given the fact that the six-digit number is randomly generated).

If the second digit is 0 to 9, the player goes into Lotto 1/300,000.

If the second digit is 0 to 4, the player goes into Lotto 1/50,000.

Lotto 1/300,000

A random number generator selects a six-digit number in the range from 100,000 to 399,999. The tickets are also randomly generated, the player does not have the option to choose the numbers.

PRIZE STRUCTURE	Probability 1 in
Division 1: Match the last five digits in the exact order drawn.	100,000
Division 2: Match the last four digits in the exact order drawn.	11,111
Division 3: Match the last three digits in the exact order drawn.	1,111

Let us visualise this lottery as if the draws were performed by using lottery ball drawing machines:

The letters **A B C D E F** represent the winning numbers.

WINNING NUMBERS					
A	B	C	D	E	F
1 2 3	0 1 2 3 4 5 6 7 8 9	0 1 2 3 4 5 6 7 8 9	0 1 2 3 4 5 6 7 8 9	0 1 2 3 4 5 6 7 8 9	0 1 2 3 4 5 6 7 8 9
U	V	W	X	Y	Z

NUMBERS IN EACH DRAW MACHINE

Calculation of Probabilities

Possible Outcomes

Because the range of the six-digit number has been defined as from 100,000 to 399,999 the number of possible outcomes is 300,000.

Favorable Outcomes and Probabilities

DIVISION 1

Match the <u>last five</u> winning numbers in the exact order drawn

U B C D E F

The letter **U** represents any of the three digits 1, 2, 3

<u>Favorable outcomes</u>
P(3,1) = 3

Probability of winning Division 1

1 in 300,000/3 = 1 in 100,000

DIVISION 2

Match the <u>last four</u> winning numbers in the exact order drawn

U V C D E F

The letter **U** represents any permutation of one number made with any of the three digits 1, 2, 3.

<u>Favorable outcomes</u>
P(3,1) = 3

The letter **V** represents any of the ten digits except **B** (V cannot be equal to **B** because **UBCDEF** is already the Division 1).

<u>Favorable outcomes</u>
P(9,1) = 9

<u>Total favorable outcomes</u>
P(3,1) x P(9,1) = 27

Probability of winning Division 2

1 in 300,000/27 = 1 in 11,111

DIVISION 3

Match the <u>last three</u> winning numbers in the exact order drawn

U B W D E F

U V W D E F

The letter U represents any permutation of one number made with any of the three digits 1, 2, 3.

$P(3,1) = 3$

<u>Favorable outcomes</u>

First row (W):

$P(3,1) \times P(9,1) = 3 \times 9 = 27$

Second row (VW):

$P(3,1) \times P(9,2) = 3 \times 81 = 243$

<u>Total favorable outcomes</u>

$27 + 243 = 270$

Probability of winning Division 3

1 in 300,000/270 = 1 in 1,111

Nonwinning arrangements

a) Matching the <u>last two</u> winning numbers

U B C X E F

U V C X E F

U B W X E F

U V W X E F

<u>Favorable outcomes</u>

First row (X):

$P(3,1) \times P(9,1) = 3 \times 9 = 27$

Second and third rows (VX and WX):

$P(3,1) \times 2[P(9,2)] = 3 \times 2 \times 81 = 486$

Fourth row (VWX):

$P(3,1) \times P(9,3) = 3 \times 729 = 2,187$

<u>Total favorable outcomes</u>

$27 + 486 + 2,187 = 2,700$

b) Matching the <u>last</u> winning number

U B C D Y F

U V C D Y F

U B W D Y F

U B C X Y F

U V W D Y F

U V C X Y F

U B W X Y F

U V W X Y F

<u>Favorable outcomes</u>

First row (Y):

P(3,1) x P(9,1) = 3 x 9 = 27

Rows second to fourth (VY, WY, and XY):

P(3,1) x 3[P(9,2)] = 3 x 3 x 81 = 729

Rows fifth to seventh (VWY, VXY, and WXY) :

P(3,1) x 3[P(9,3)] = 3 x 3 x 729 = 6,561

Eighth row (VWXY):

P(3,1) x P(9,4) = 3 x 6,561 = 19,683

<u>Total favorable outcomes</u>

27 + 729 + 6,561 + 19,683 = 27,000

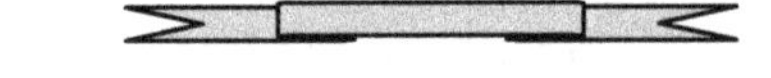

c) Not matching any winning number

Different nonwinning arrangements:

1)

U B C D E Z

<u>Favorable outcomes</u>

P(3,1) x P(9,1) = 3 x 9 = 27

2)

U B C D Y Z

U B C X E Z

U B W D E Z

U V C D E Z

<u>Favorable outcomes</u>

P(3,1) x 4[P(9,2)] = 3 x 4 x 81 = 972

3)

U B C X Y Z

U B W D Y Z

U B W X E Z

U V C D Y Z

U V C X E Z

U V W D E Z

Favorable outcomes

P(3,1) x 6[P(9,3)] = 3 x 6 x 729 = 13,122

4)

U B W X Y Z

U V C X Y Z

U V W D Y Z

U V W X E Z

Favorable outcomes

P(3,1) x 4[P(9,4)] = 3 x 4 x 6,561 = 78,732

5)

U V W X Y Z

Favorable outcomes

P(3,1) x P(9,5) = 3 x 59,049 = 177,147

Total favorable outcomes:

27 + 972 + 13,122 + 78,732 + 177,147 = 270,000

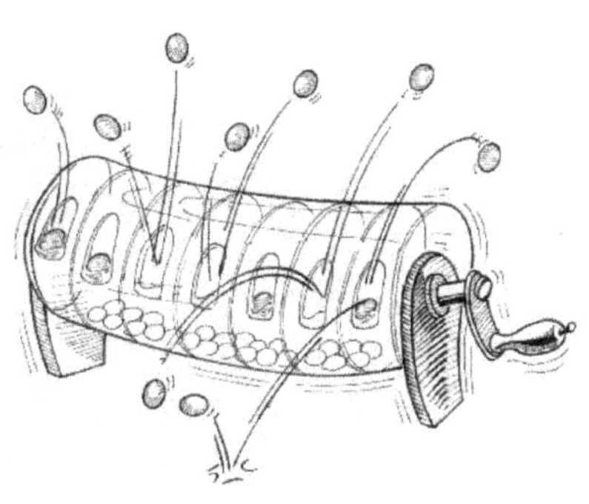

Summary

PRIZE DIVISION							U (First number)	N (Numbers UVWXYZ)	U × N	FAVORABLE OUTCOMES
DIVISION 1	U	B	C	D	E	F	$P(3,1) = 3$		3	3
DIVISION 2	U	V	C	D	E	F	$P(3,1) = 3$	$P(9,1) = 9$	27	27
DIVISION 3	U	B	W	D	E	F	$P(3,1) = 3$	$P(9,1) = 9$	27	270
	U	V	W	D	E	F	$P(3,1) = 3$	$P(9,2) = 81$	243	
Matching the last two winning numbers	U	B	C	X	E	F	$P(3,1) = 3$	$P(9,1) = 9$	27	2,700
	U	V	C	X	E	F	$P(3,1) = 3$	$P(9,2) = 81$	243	
	U	B	W	X	E	F	$P(3,1) = 3$	$P(9,2) = 81$	243	
	U	V	W	X	E	F	$P(3,1) = 3$	$P(9,3) = 729$	2,187	
Matching the last winning number	U	B	C	D	Y	F	$P(3,1) = 3$	$P(9,1) = 9$	27	27,000
	U	V	C	D	Y	F	$P(3,1) = 3$	$P(9,2) = 81$	243	
	U	B	W	D	Y	F	$P(3,1) = 3$	$P(9,2) = 81$	243	
	U	B	C	X	Y	F	$P(3,1) = 3$	$P(9,2) = 81$	243	
	U	V	W	D	Y	F	$P(3,1) = 3$	$P(9,3) = 729$	2,187	
	U	V	C	X	Y	F	$P(3,1) = 3$	$P(9,3) = 729$	2,187	
	U	B	W	X	Y	F	$P(3,1) = 3$	$P(9,3) = 729$	2,187	
	U	V	W	X	Y	F	$P(3,1) = 3$	$P(9,4) = 6,561$	19,683	
Matching no winning numbers	U	B	C	D	E	Z	$P(3,1) = 3$	$P(9,1) = 9$	27	270,000
	U	B	C	D	Y	Z	$P(3,1) = 3$	$P(9,2) = 81$	243	
	U	B	C	X	E	Z	$P(3,1) = 3$	$P(9,2) = 81$	243	
	U	B	W	D	E	Z	$P(3,1) = 3$	$P(9,2) = 81$	243	
	U	V	C	D	E	Z	$P(3,1) = 3$	$P(9,2) = 81$	243	
	U	B	C	X	Y	Z	$P(3,1) = 3$	$P(9,3) = 729$	2,187	
	U	B	W	D	Y	Z	$P(3,1) = 3$	$P(9,3) = 729$	2,187	
	U	B	W	X	E	Z	$P(3,1) = 3$	$P(9,3) = 729$	2,187	
	U	V	C	D	Y	Z	$P(3,1) = 3$	$P(9,3) = 729$	2,187	
	U	V	C	X	E	Z	$P(3,1) = 3$	$P(9,3) = 729$	2,187	
	U	V	W	D	E	Z	$P(3,1) = 3$	$P(9,3) = 729$	2,187	
	U	B	W	X	Y	Z	$P(3,1) = 3$	$P(9,4) = 6,561$	19,683	
	U	V	C	X	Y	Z	$P(3,1) = 3$	$P(9,4) = 6,561$	19,683	
	U	V	W	D	Y	Z	$P(3,1) = 3$	$P(9,4) = 6,561$	19,683	
	U	V	W	X	E	Z	$P(3,1) = 3$	$P(9,4) = 6,561$	19,683	
	U	V	W	X	Y	Z	$P(3,1) = 3$	$P(9,5) = 59,049$	177,147	
POSSIBLE OUTCOMES										300,000

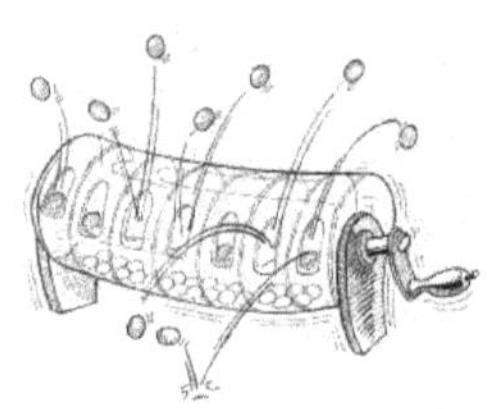

Lotto 1/50,000

A random number generator selects a six-digit number in the range from 400,000 to 449,999. The tickets are also randomly generated, the player does not have the option to choose the numbers.

PRIZE STRUCTURE		Probability 1 in
Division 1:	Match the last five digits in the exact order drawn.	50,000
Division 2:	Match the last four digits in the exact order drawn.	12,500
Division 3:	Match the last three digits in the exact order drawn.	1,111

Let us visualise this lottery as if the draws were performed by using lottery ball drawing machines:

The letters **A B C D E F** represent the winning numbers.

WINNING NUMBERS					
A	B	C	D	E	F
4	0 1 2 3 4 5 6 7 8 9	0 1 2 3 4 5 6 7 8 9	0 1 2 3 4 5 6 7 8 9	0 1 2 3 4 5 6 7 8 9	0 1 2 3 4 5 6 7 8 9
U	V	W	X	Y	Z
NUMBERS IN EACH DRAW MACHINE					

Calculation of Probabilities

Possible Outcomes

Because the range of the six-digit number has been defined as from 400,000 to 449,999 the number of possible outcomes is 50,000.

Favorable Outcomes and Probabilities

DIVISION 1
Match the <u>last five</u> winning numbers in the exact order drawn

U B C D E F

The letter U represents the digit 4

<u>Favorable outcomes</u>
P(1,1) = 1

Probability of winning Division 1

1 in 50,000/1 = 1 in 50,000

DIVISION 2
Match the <u>last four</u> winning numbers in the exact order drawn

U V C D E F

The letter U represents the digit 4.
The letter V represents any of the <u>four</u> digits 0 to 4 except **B** (V cannot be equal to **B** because **UBCDEF** is already the Division 1).

<u>Favorable outcomes</u>
P(4,1) = 4

Probability of winning Division 2

1 in 50,000/4 = 1 in 12,500

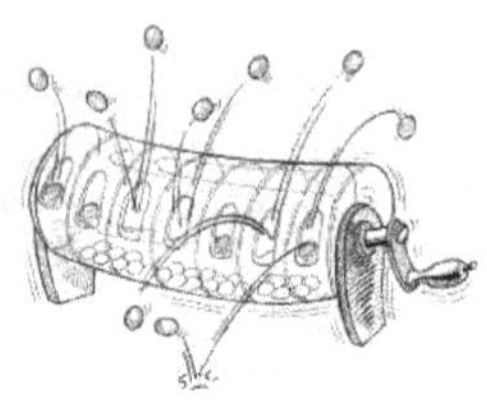

DIVISION 3

Match the <u>last three</u> winning numbers in the exact order drawn

U V W D E F

The letter U represents the digit 4.

The letter V represents either of the <u>five</u> digits 0 to 4.

The letter W represents either of the <u>nine</u> digits 0 to 9 except **C** (W cannot be equal to **C** because UV**CDEF** is already the Division 2).

<u>Favorable outcomes</u>

$P(5,1) \times P(9,1) = 5 \times 9 = 45$

Probability of winning Division 3

1 in 50,000/45 = 1 in 1,111

Nonwinning arrangements

a) Matching the <u>last two</u> winning numbers

U V C X E F

U V W X E F

The letter V represents either of the <u>five</u> digits 0 to 4.

Favorable outcomes:

First row:

$P(5,1) \times P(9,1) = 5 \times 9 = 45$

Second row:

$P(5,1) \times P(9,2) = 5 \times 81 = 405$

Total favorable outcomes

45 + 405 = 450

b) Matching the <u>last</u> winning number

U V C D Y F

U V W D Y F

U V C X Y F

U V W X Y F

The letter V represents either of the <u>five</u> digits 0 to 4.

Favorable outcomes:

First row:

$P(5,1) \times P(9,1) = 5 \times 9 = 45$

Rows second to third:
2[P(5,1) x P(9,2)] = 2 x 5 x 81 = 810

Fourth row:
P(5,1) x P(9,3) = 5 x 729 = 3,645

<u>Total favorable outcomes</u>
45 + 810 + 3,645 = 4,500

c) Not matching any winning number
(The letter U represents the digit 4 in all cases, and the letter V represents either of the <u>five</u> digits 0 to 4 in all cases).

Different possible arrangements:

1)

U V C D E Z

<u>Favorable outcomes</u>
P(5,1) x P(9,1) = 5 x 9 = 45

2)

U V C D Y Z

U V C X E Z

U V W D E Z

<u>Favorable outcomes</u>
3[P(5,1) x P(9,2)] = 3 x 5 x 81 = 1,215

3)

U V C X Y Z

U V W D Y Z

U V W X E Z

<u>Favorable outcomes</u>
3[P(5,1) x P(9,3)] = 3 x 5 x 729 = 10,935

4)

U V W X Y Z

<u>Favorable outcomes</u>
P(5,1) x P(9,4) = 5 x 6,561 = 32,805

<u>Total favorable outcomes:</u>
45 + 1,215 + 10,935 + 32,805 = 45,000

Summary

PRIZE DIVISION								V (Second number)	N (Numbers WXYZ)	V × N	FAVORABLE OUTCOMES
DIVISION 1	U	B	C	D	E	F		$P(4,0) = 1$	$P(9,0) = 1$	1	1
DIVISION 2	U	V	C	D	E	F		$P(4,1) = 4$	$P(9,0) = 1$	4	4
DIVISION 3	U	V	W	D	E	F		$P(5,1) = 5$	$P(9,1) = 9$	45	45
Non-winning a)	U	V	C	X	E	F		$P(5,1) = 5$	$P(9,1) = 9$	45	450
	U	V	W	X	E	F		$P(5,1) = 5$	$P(9,2) = 81$	405	
Non-winning b)	U	V	C	D	Y	F		$P(5,1) = 5$	$P(9,1) = 9$	45	4,500
	U	V	W	D	Y	F		$P(5,1) = 5$	$P(9,2) = 81$	405	
	U	V	C	X	Y	F		$P(5,1) = 5$	$P(9,2) = 81$	405	
	U	V	W	X	Y	F		$P(5,1) = 5$	$P(9,3) = 729$	3,645	
Non-winning c)	U	V	C	D	E	Z		$P(5,1) = 5$	$P(9,1) = 9$	45	45,000
	U	V	C	D	Y	Z		$P(5,1) = 5$	$P(9,2) = 81$	405	
	U	V	C	X	E	Z		$P(5,1) = 5$	$P(9,2) = 81$	405	
	U	V	W	D	E	Z		$P(5,1) = 5$	$P(9,2) = 81$	405	
	U	V	C	X	Y	Z		$P(5,1) = 5$	$P(9,3) = 729$	3,645	
	U	V	W	D	Y	Z		$P(5,1) = 5$	$P(9,3) = 729$	3,645	
	U	V	W	X	E	Z		$P(5,1) = 5$	$P(9,3) = 729$	3,645	
	U	V	W	X	Y	Z		$P(5,1) = 5$	$P(9,4) = 6,561$	32,805	
							POSSIBLE OUTCOMES =				50,000

7.4 Is Official Lottery Information Always Accurate?

It is not unusual to find inaccuracies in lottery websites. The following are two real life examples.

Lotto 6/45 — Two Supplementary Numbers

A lottery ball drawing machine contains forty-five balls numbered from 1 to 45. Six balls are drawn without replacement. Those are the winning numbers.

Two more balls are drawn without replacement. Those are the supplementary numbers.

PRIZE STRUCTURE	Probabilities (For 12 games)
	1 in
Division 1: Match all six winning numbers.	678,755
Division 2: Match five winning numbers and *one or two supplementary numbers.*	56,563
Division 3: Match five winning numbers.	3,057
Division 4: Match four winning numbers.	61
Division 5: Match three winning numbers and one or two supplementary numbers.	25
Division 6: Match one or two winning numbers and two supplementary numbers.	12

Division 2 is incorrectly defined. Let us see why:

Division 2

Division 2 - Match five winning numbers and one or two supplementary numbers.

Winning numbers

Supplementary number

A B C D E

G

The letters **A B C D E** represent any combination of five numbers made with the six winning numbers.

The letter **G** represents any combination of one number made with the two supplementary numbers.

It is not possible to match five winning numbers and two supplementary numbers, because this will require selecting seven numbers in the game panel, which is not allowed according to the definition of this lottery.

The information with Division 2 wrongly defined was published in 2016 by the following lotteries:

Monday & Wednesday Lotto, and Saturday Lotto — New South Wales (Australia). Monday & Wednesday Gold Lotto, and Saturday Gold Lotto — Queensland (Australia).
Monday & Wednesday X Lotto, and Saturday X Lotto — South Australia (Australia).
TattsLotto — Victoria, Tasmania, Northern Territory (Australia).

Lotto 7/45 — Two Supplementary Numbers

A lottery ball drawing machine contains forty-five balls numbered 1 to 45. Seven balls are drawn without replacement. Those are the winning numbers.

Two more balls are drawn without replacement. Those are the supplementary numbers.

PRIZE STRUCTURE	Probabilities (For 12 games) 1 in
Division 1: Match all seven winning numbers.	3,781,635
Division 2: Match six winning numbers and *one or two supplementary numbers.*	270,117
Division 3: Match six winning numbers.	15,007
Division 4: Match five winning numbers and one or two supplementary numbers.	2,467

Division 5: Match five winning numbers.	286
Division 6: Match four winning numbers.	13
Division 7: Match three winning numbers and one or two supplementary numbers.	7

Division 2 is incorrectly defined. Let us see why:

Division 2

Division 2 - Match five winning numbers and one or two supplementary numbers.	
Winning numbers	Supplementary number
A B C D E F	**G**

The letters **A B C D E F** represent any combination of six numbers made with the seven winning numbers.

The letter **G** represents any combination of one number made with the two supplementary numbers.

It is not possible to match six winning numbers and two supplementary numbers, because this will require selecting eight numbers in the game panel, which is not allowed according to the definition of this lottery.

The information with Division 2 wrongly defined was published in 2016 by the following lotteries:

Oz Lotto—New South Wales, Queensland, Victoria, South Australia, Tasmania, Northern Territory (Australia).

The following explanations (text highlighted in grey) were obtained from Tatts Group Customer Support:

System entries are a ticket that plays more numbers in a game than the standard. On any ticket you will always be paid the highest number combination achieved in a game.

For example:

1) A Saturday Standard Entry will have 6 numbers per game, if you were to match 5 Winning Numbers + 1 Supplementary, this will be the highest possible combination for Division 2 to pay, even if a second supplementary matched in the same game. It is important to note here that a second supplementary number would provide no extra dividend as presented on our results.

2) A Saturday System 9 Entry will have 9 numbers per game. This increases the number of combinations possible, including 5 Winning Numbers + 2 Supplementaries. Prizes on a System Entry can be multiples of divisions.

As we have multiple game types available, we show some of these basic combinations in our game results to accommodate all of our players.

You may find further clarification in the 'How to Play Systems Guide', which shows the valid number combinations for system entries. I have attached a copy to this email for your convenience.

Customer Support

We acknowledge that the odds of winning a division 2 prize in Lotto with a standard entry where you match five winning numbers with one supplementary number are different to the odds of winning a division 2 prize with a system entry where it is possible to match five winning numbers plus one or two supplementary numbers. This is consistent with the relevant Lotto Game Rules which state that –

"A Prize of an amount equal to 4.5% of the Prize Pool shall be payable in respect of any Entry or Syndicate Entry which contains five (5) but not more than five (5) of the six (6) Winning Numbers together with one (1) or both of the Supplementary Numbers."

The example given within the "Let's Play" flyer are the odds for winning a division 2 prize with a 12-game standard entry and this is qualified with an asterisk to draw attention to this. On the website the odds are labelled as relating to either a single standard game or 12 game standard entry which is different to a system entry as you have correctly noted in your email.

Customer Support

JAIME AGUIRRE

<u>Author's response:</u>

There are two types of misperceptions here:

1.   Numbers and games:

Taking a 6/45 lottery as an example:
The numbers are 1 to 45.
A game is any set of 6 numbers made with any six of the 45 numbers.

2.   Numbers played and winning numbers:

Taking a 6/45 lottery as an example, where the player chooses to play 8 numbers.
The numbers played are the 8 numbers chosen by the player, being any eight numbers from 1 to 45.

The winning numbers are the 6 numbers drawn by the lottery draw machine.

Playing a system entry does not change the definition of the lottery and its prize structure. A system is simply a method of playing many games without marking every single game in the panels. A person can play the equivalent to a system without playing a system entry, but marking every possible combination, and therefore the Division 2 prize must be the same for him as for the player marking the same numbers in a system entry.

Let us see an example:

Player A selects eight numbers in a system entry. The numbers chosen are 1, 2, 3, 4, 5, 6, 7, and 8. He only must mark these numbers in a panel, and the computer enters all possible combinations for him.
This is equivalent to Player A marking 28 standard games, as follows:

1	2	3	4	5	6
1	2	3	4	5	7
1	2	3	4	5	8
1	2	3	4	6	7
1	2	3	4	6	8

1	2	3	4	7	8
1	2	3	5	6	7
1	2	3	5	6	8
1	2	3	5	7	8
1	2	3	6	7	8
1	2	4	5	6	7
1	2	4	5	6	8
1	2	4	5	7	8
1	2	4	6	7	8
1	2	5	6	7	8
1	3	4	5	6	7
1	3	4	5	6	8
1	3	4	5	7	8
1	3	4	6	7	8
1	3	5	6	7	8
1	4	5	6	7	8
2	3	4	5	6	7
2	3	4	5	6	8
2	3	4	5	7	8
2	3	4	6	7	8
2	3	5	6	7	8
2	4	5	6	7	8
3	4	5	6	7	8

Player B also selects the same eight numbers, but he decides to play only three standard games. The games are:

1	2	3	4	5	6
1	2	3	4	5	7
1	2	3	4	5	8

It is particularly important to note that both players A and B are playing exactly the same lottery, and therefore the prize definition must be the same for each one.

Suppose the winning numbers are 1, 2, 3, 4, 5, and 45.

Suppose the supplementary numbers are 7 and 8.

The matching of these numbers can be defined in either of two ways:

1.   <u>In relation to the set of eight numbers played</u>

Both players matched five winning numbers (1, 2, 3, 4 and 5) and two supplementary numbers (7 and 8).

2.   <u>In relation to the set of six winning numbers</u>

Both players matched five winning numbers (1, 2, 3, 4 and 5) and either supplementary number (7 or 8). In this case you cannot say that they matched five winning numbers and two supplementary numbers, because the set must comprise only six numbers, as defined by the lottery structure.

The table defining the number of prizes is applicable to both players. Each player won two Division 2 prizes. The only problem with this table is that the column "Winning Combinations" should say "Numbers matched in the set of numbers played".

Lotto									
Win. Comb.	**Div.**	**SYSTEM**							
		7	**8**	**9**	**10**	**11**	**12**	**13**	**14**
5 W and 2 S	**2**	2	2	2	2	2	2	2	2

Win. Comb. = Winning Combination
Div. = Division
W = Winning numbers
S = Supplementary numbers

If the prize structure were different for a person playing systems, it should be stated. Nevertheless, this would be a non-sense, because playing systems is simply a method to play many games without marking every single game in the panels, therefore the prize structure should be unique.

When you read the brochure on Saturday Lotto, it says:

"What you need to win Saturday Lotto"

"Division 2 - Match five winning numbers and one or two supplementary numbers"

It is not stated that this definition applies in relation to the set of numbers played. And because every lottery in the world defines its prize structure in relation to the set of winning numbers, it sounds as if the above definition of Division 2 also refers to the set of winning numbers, and so it looks wrong, because the set of winning numbers is six, not seven.

To clarify why division 2 says 5 winning numbers plus one or two supplementary numbers or in Oz Lotto it says 6 winning numbers plus one or two supplementary numbers.
In the standard games where 6 numbers are selected for purchasing it is only possible to get 5 winning numbers plus a supplementary number to win division 2.
We also offer players System entry tickets where you can select between 7 to 20 numbers. A customer chooses say 14 numbers, their 2nd division can be made of 5 winning numbers plus two supplementary numbers."
Customer Support

Author's response:

Even when playing systems, the second division cannot be made of 5 winning numbers plus two supplementary numbers. It can be made of 5 winning numbers plus either supplementary number. A customer playing systems could match the two supplementary numbers, but even so the division 2 cannot be made of 5 winning numbers plus two supplementary numbers.

For example:

In Saturday Lotto Draw 3613 (5 March 2016) the winning numbers were 2 11 20 26 39 45. And the supplementary numbers were 19 33.

If a customer played any system including the numbers 2 11 19 20 26 33 and 39, he matched 5 winning numbers and the 2 supplementary numbers, but his 2nd division cannot be made of the numbers 2 11 19 20 26 33 and 39, because that is a seven number panel, which is not allowed. He simply won two Division 2 prizes:

Division 2 - Prize 1:

2 11 20 26 39 19

Division 2 - Prize 2:

2 11 20 26 39 33

When playing the relevant games, there are different types of entries that can be purchased. These include Standard entries and System entries.

From the Rules of Authorised Lotteries (see page 7) they are defined as:

Definition of a standard entry - is a selection (however made) of a single number or the amount of numbers as prescribed in the Schedule for each Authorised Lottery, if applicable

Definition of a system entry - means (unless otherwise defined in a Schedule) an Entry:

(a) arising from the selection of numbers in excess of those prescribed for a Standard Entry for the relevant Authorised Lottery; and

(b) where the Tatts lottery system combines all possible Standard Entry number combinations for the relevant Authorised Lottery from the selected numbers

Using Oz Lotto as an example, division of prizes – Division Two (page 19) - 6 of the 7 Winning Numbers plus either of the Supplementary Numbers in any one Standard Entry

On a system entry where 8 or more numbers are played, is it possible to have for a 2nd division win, 6 numbers and 2 supplementary numbers.

On our website the winning combinations for a 2nd division covers all entries and includes Standard and System entries.

Division 2 - 6 winning numbers +1 or 2 supplementary number

Should you have any further queries, please email us, telephone our Contact Centre on 131 868, or visit our website http://tatts.com/nswlotteries Customer Support

<u>Author's response:</u>

<u>Calculating the odds of winning **Division 2**</u>
<u>Monday & Wednesday Lotto, Saturday Lotto</u>

Division 2

Division 2 - Match five winning numbers and **<u>one supplementary number.</u>**		
Winning numbers		Supplementary number
A B C D E		G

The letters **A B C D E** represent any combination of **<u>five numbers</u>** made with the six winning numbers:

$$C(6,5) = 6$$

The letter **G** represents any combination of **<u>one number</u>** made with the two supplementary numbers:

$$C(2,1) = 2$$

Favorable outcomes = 6 × 2 = 12

Odds = 12 in 8,145,060 = 1 in 678,755

<u>And for twelve games:</u>

Odds = 1 in 678,755/12 = **1 in 56,563**

Which is the same as stated in the Tatts brochure:

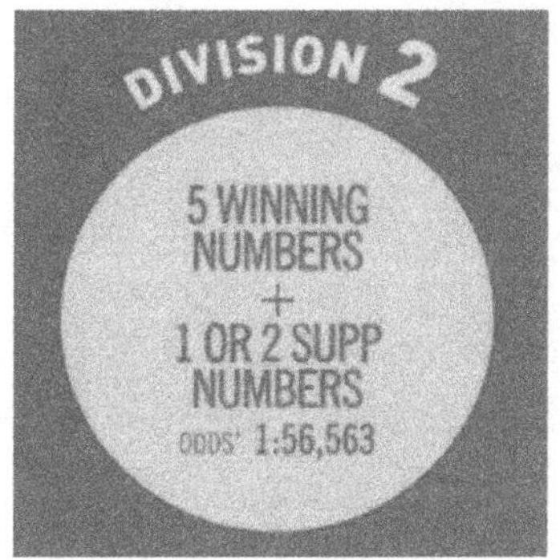

How can you obtain the same odds using any combination of **<u>two numbers</u>** made with the two supplementary numbers? It is impossible! Therefore, the odds correspond only to the definition "Match five winning numbers and <u>one supplementary number</u>."

At the date of this publication (August 2020) the actual definition has been updated in the NSW Lotteries website, as follows:

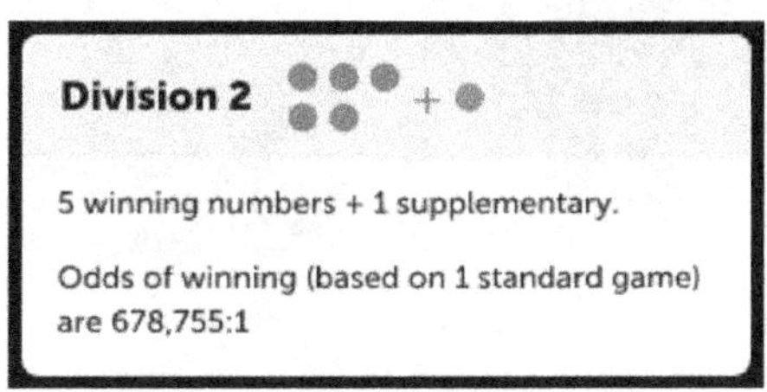

<u>Calculating the odds of winning **Division 2**</u>
<u>Oz Lotto</u>

Division 2

Division 2 - Match six winning numbers and **<u>one supplementary number.</u>**		
Winning numbers		Supplementary number
A B C D E F		**G**

The letters **A B C D E F** represent any combination of **<u>six numbers</u>** made with the seven winning numbers:

$$C(7,6) = 7$$

The letter **G** represents any combination of **<u>one number</u>** made with the two supplementary numbers:

$$C(2,1) = 2$$

Favorable outcomes = 7 × 2 = 14

Odds = 14 in 45,379,620 = 1 in 3,241,401

<u>And for twelve games:</u>

Odds = 1 in 3,241,401/12 = **1 in 270,117**

Which is the same as stated in the Tatts brochure:

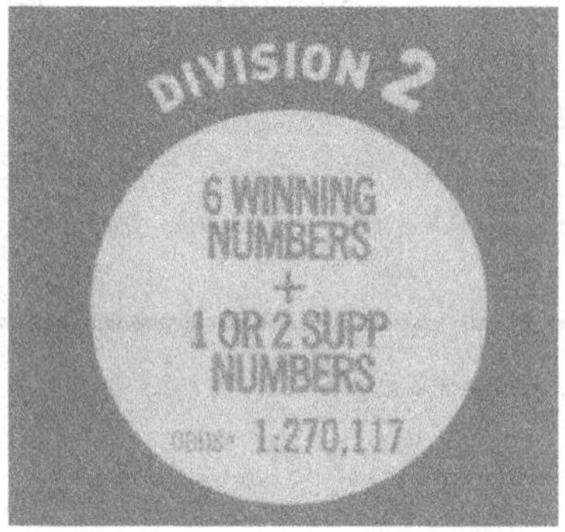

How can you obtain the same odds using any combination of **two numbers** made with the two supplementary numbers? It is impossible! Therefore, the odds correspond only to the definition "Match six winning numbers and <u>one supplementary number.</u>"

At the date of this publication (August 2020) the actual definition has been updated in the NSW Lotteries website, as follows:

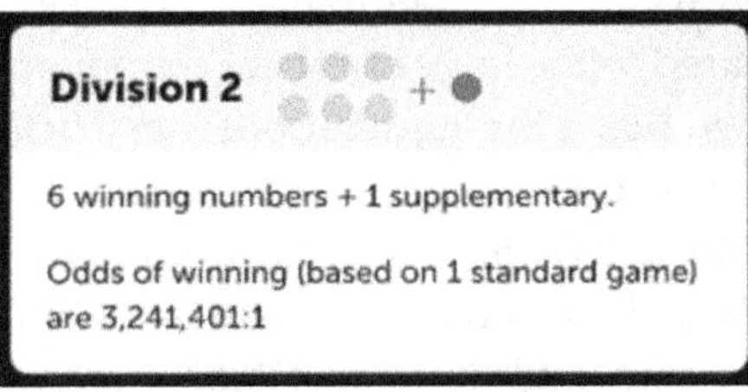

7.5 The Tip of the Iceberg

As a final note to this study on the different types of lotteries, it is worth always keeping in mind that, as a rule, the big prizes are hard to catch. The probability of winning the Division 1 in any lottery is extremely low, and the bigger the prize, the lower the probability.

The prize divisions, or favorable outcomes, are always a very small amount compared to the huge number of unfavorable outcomes, and it must be that way so that a big number of lottery ticket buyers build up the amount of money required to pay those prizes.

Take for example Lotto 7/45—Two supplementary numbers. The winning combinations (Divisions 1 through 7) add up to 832,140. The nonwinning combinations add up to 45,379,620. This means that only 1.8 per-cent of all possible outcomes are winners (See figures 7.2 and 7.3).

FIGURE 7.2 The Tip of the Iceberg

Table 7.10 shows some examples of how small the favorable outcomes (winning combinations) are when compared to the total possible outcomes (total possible combinations—permutations in the case of Lotto Strike—).

But do not jump to any happy conclusion yet: do not forget that most of the prizes are exceedingly small, and sometimes the amount you win does not even recoup the money you spent on buying tickets. In brief, if the overall chances of winning a prize are good, that is because there are many small prizes included in the favorable outcomes. These insignificant prizes make up the bulk of the winning combinations, and thus enhance the overall chances of winning a prize in a particular lottery.

TABLE 7.10 Winning outcomes as percentages of the total possible outcomes

Lottery	Winning Outcomes	Total Possible Outcomes	Percent	Reference
	WO	TPO	$100 \times \dfrac{WO}{TPO}$	
5/28	20,356	98,280	20.7	Page 76
6/48 Plus 1/5	9,608,590	61,357,560	15.7	Page 230
6/53	3,016,273	22,957,480	13.1	Page 125
5/52 Plus 1/10	2,698,374	25,989,600	10.4	Page 212
5/50 Plus 2/12	10,778,691	139,838,160	7.7	Page 210
5/70 plus 1/25	12,610,038	302,575,350	4.2	Page 220
5/69 Plus 1/26	11,750,538	292,201,338	4.0	Page 218
6/45 (2 suppl. numbers) Prize Str. A	194,130	8,145,060	2.4	Page 107
7/35 Plus 1/20	3,057,865	134,490,400	2.3	Page 240
7/47 (3 suppl. numbers)	1,249,536	62,891,499	2.0	Page 152
7/45 (2 suppl. numbers)	832,140	45,379,620	1.8	Page 150
6/45 (2 suppl. numbers) Prize Str. B	95,340	8,145,060	1.2	Page 109
6/40 (1 suppl. number) Plus 1/10	128,300	38,383,800	0.3	Page 222

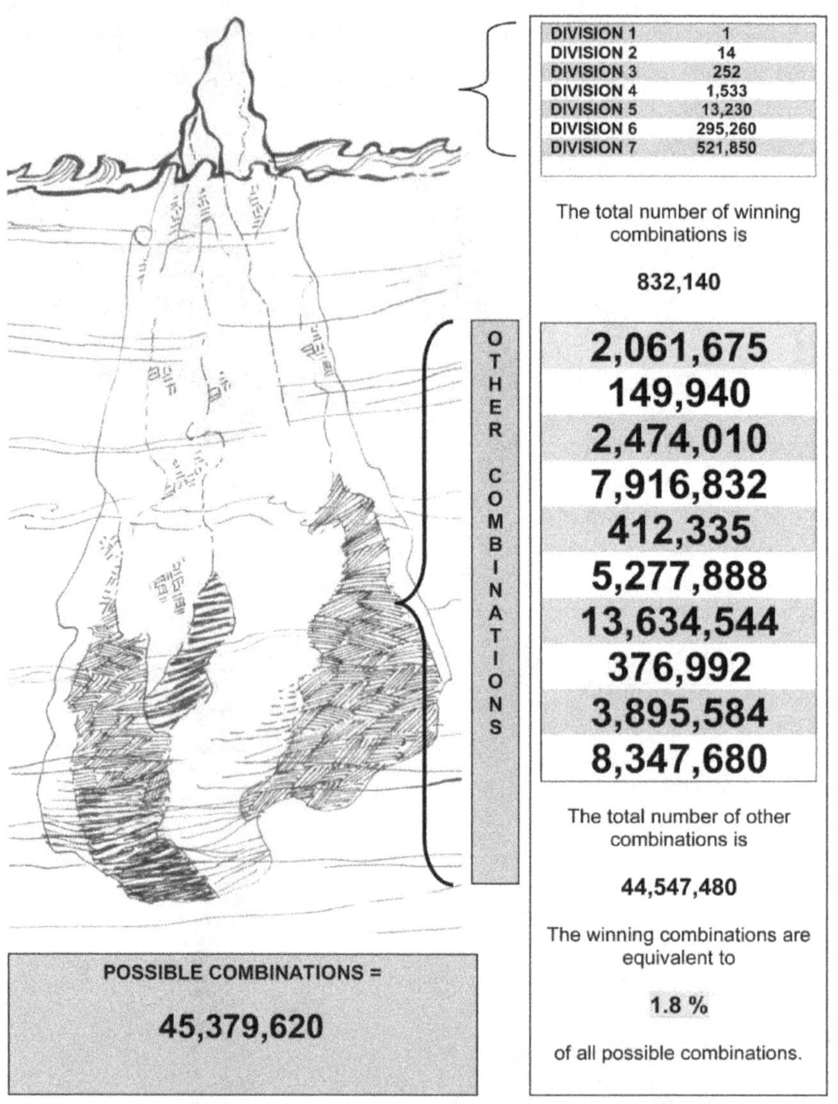

FIGURE 7.3 The Lotto Iceberg:
Lotto 7/45 — Two supplementary numbers

7.6 If My Calculations are Correct...

How can we be certain that the probability calculations are correct?

One method is to work out all possible outcomes (favorable and unfavorable). The result must be the same as the total possible outcomes. Let us see some examples:

5/10 — Exact order

All five	$1 \times P(9,0)$	$= 1 \times 1$	$=$	1
				1
First (or last) four	$2 \times P(9,1)$	$= 2 \times 9$	$=$	18
				18
First (or last) three	$2 \times P(9,1)$	$= 2 \times 9$	$=$	18
	$2 \times P(9,2)$	$= 2 \times 81$	$=$	162
				180
First (or last) two	$1 \times P(9,1)$	$= 1 \times 9$	$=$	9
	$4 \times P(9,2)$	$= 4 \times 81$	$=$	324
	$2 \times P(9,3)$	$= 2 \times 729$	$=$	1,458
				1,791
First (or last) one	$3 \times P(9,2)$	$= 3 \times 81$	$=$	243
	$5 \times P(9,3)$	$= 5 \times 729$	$=$	3,645
	$2 \times P(9,4)$	$= 2 \times 6,561$	$=$	13,122
				17,010
Other permutations	$1 \times P(9,2)$	$= 1 \times 81$	$=$	81
	$3 \times P(9,3)$	$= 3 \times 729$	$=$	2,187
	$3 \times P(9,4)$	$= 3 \times 6,561$	$=$	19,683
	$1 \times P(9,5)$	$= 1 \times 59,049$	$=$	59,049
				81,000
	Total possible outcomes:			100,000

$$P(n,r) = n^r = P(10,5) = 10^5 = 100,000$$

7/35 (4 supplementary numbers)

W		S		N		W	S	N			
$_7C_7$	×	$_4C_0$	×	$_{24}C_0$	=	1 ×	1 ×	1	=	1	1 DIV 1
$_7C_6$	×	$_4C_1$	×	$_{24}C_0$	=	7 ×	4 ×	1	=	28	28 DIV 2
$_7C_6$	×	$_4C_0$	×	$_{24}C_1$	=	7 ×	1 ×	24	=	168	168 DIV 3
$_7C_5$	×	$_4C_2$	×	$_{24}C_0$	=	21 ×	6 ×	1	=	126	
$_7C_5$	×	$_4C_1$	×	$_{24}C_1$	=	21 ×	4 ×	24	=	2,016	7,938 DIV 4
$_7C_5$	×	$_4C_0$	×	$_{24}C_2$	=	21 ×	1 ×	276	=	5,796	
$_7C_4$	×	$_4C_3$	×	$_{24}C_0$	=	35 ×	4 ×	1	=	140	
$_7C_4$	×	$_4C_2$	×	$_{24}C_1$	=	35 ×	6 ×	24	=	5,040	
$_7C_4$	×	$_4C_1$	×	$_{24}C_2$	=	35 ×	4 ×	276	=	38,640	114,660 DIV 5
$_7C_4$	×	$_4C_0$	×	$_{24}C_3$	=	35 ×	1 ×	2,024	=	70,840	
$_7C_3$	×	$_4C_4$	×	$_{24}C_0$	=	35 ×	1 ×	1	=	35	
$_7C_3$	×	$_4C_3$	×	$_{24}C_1$	=	35 ×	4 ×	24	=	3,360	
$_7C_3$	×	$_4C_2$	×	$_{24}C_2$	=	35 ×	6 ×	276	=	57,960	716,625 DIV 6
$_7C_3$	×	$_4C_1$	×	$_{24}C_3$	=	35 ×	4 ×	2,024	=	283,360	
$_7C_3$	×	$_4C_0$	×	$_{24}C_4$	=	35 ×	1 ×	10,626	=	371,910	
$_7C_2$	×	$_4C_4$	×	$_{24}C_1$	=	21 ×	1 ×	24	=	504	
$_7C_2$	×	$_4C_3$	×	$_{24}C_2$	=	21 ×	4 ×	276	=	23,184	
$_7C_2$	×	$_4C_2$	×	$_{24}C_3$	=	21 ×	6 ×	2,024	=	255,024	2,063,880
$_7C_2$	×	$_4C_1$	×	$_{24}C_4$	=	21 ×	4 ×	10,626	=	892,584	
$_7C_2$	×	$_4C_0$	×	$_{24}C_5$	=	21 ×	1 ×	42,504	=	892,584	
$_7C_1$	×	$_4C_4$	×	$_{24}C_2$	=	7 ×	1 ×	276	=	1,932	
$_7C_1$	×	$_4C_3$	×	$_{24}C_3$	=	7 ×	4 ×	2,024	=	56,672	
$_7C_1$	×	$_4C_2$	×	$_{24}C_4$	=	7 ×	6 ×	10,626	=	446,292	2,637,180
$_7C_1$	×	$_4C_1$	×	$_{24}C_5$	=	7 ×	4 ×	42,504	=	1,190,112	
$_7C_1$	×	$_4C_0$	×	$_{24}C_6$	=	7 ×	1 ×	134,596	=	942,172	
$_7C_0$	×	$_4C_4$	×	$_{24}C_3$	=	1 ×	1 ×	2,024	=	2,024	
$_7C_0$	×	$_4C_3$	×	$_{24}C_4$	=	1 ×	4 ×	10,626	=	42,504	
$_7C_0$	×	$_4C_2$	×	$_{24}C_5$	=	1 ×	6 ×	42,504	=	255,024	1,184,040
$_7C_0$	×	$_4C_1$	×	$_{24}C_6$	=	1 ×	4 ×	134,596	=	538,384	
$_7C_0$	×	$_4C_0$	×	$_{24}C_7$	=	1 ×	1 ×	346,104	=	346,104	

TOTAL POSSIBLE OUTCOMES = 6,724,520   6,724,520

$$_{24}C_7 = 6,724,520$$

5/43 Plus 1/16

MACHINE A		MACHINE B					OUTCOMES	
5 C 5 x 38 C 0	x 1 C 1 x 15 C 0	= 1 x	1 x 1 x 1 =				1	DIV 1
5 C 5 x 38 C 0	x 1 C 0 x 15 C 1	= 1 x	1 x 1 x 15 =				15	DIV 2
5 C 4 x 38 C 1	x 1 C 1 x 15 C 0	= 5 x	38 x 1 x 1 =				190	DIV 3
5 C 4 x 38 C 1	x 1 C 0 x 15 C 1	= 5 x	38 x 1 x 15 =				2,850	DIV 4
5 C 3 x 38 C 2	x 1 C 1 x 15 C 0	= 10 x	703 x 1 x 1 =				7,030	DIV 5
5 C 3 x 38 C 2	x 1 C 0 x 15 C 1	= 10 x	703 x 1 x 15 =				105,450	DIV 6
5 C 2 x 38 C 3	x 1 C 1 x 15 C 0	= 10 x	8,436 x 1 x 1 =				84,360	DIV 7
5 C 2 x 38 C 3	x 1 C 0 x 15 C 1	= 10 x	8,436 x 1 x 15 =				1,265,400	
5 C 1 x 38 C 4	x 1 C 1 x 15 C 0	= 5 x	73,815 x 1 x 1 =				369,075	DIV 8
5 C 1 x 38 C 4	x 1 C 0 x 15 C 1	= 5 x	73,815 x 1 x 15 =				5,536,125	
5 C 0 x 38 C 5	x 1 C 1 x 15 C 0	= 1 x	501,942 x 1 x 1 =				501,942	DIV 9
5 C 0 x 38 C 5	x 1 C 0 x 15 C 1	= 1 x	501,942 x 1 x 15 =				7,529,130	

TOTAL POSSIBLE OUTCOMES: 15,401,568

$C_{(43,5)} \times C_{(16,1)} = 962,598 \times 16 = 15,401,568$

5/49 Plus 1/13

MACHINE A		MACHINE B					OUTCOMES	
5 C 5 x 44 C 0	x 1 C 1 x 12 C 0	= 1 x	1 x 1 x 1 =				1	DIV 1
5 C 5 x 44 C 0	x 1 C 0 x 12 C 1	= 1 x	1 x 1 x 12 =				12	DIV 2
5 C 4 x 44 C 1	x 1 C 1 x 12 C 0	= 5 x	44 x 1 x 1 =				220	
5 C 4 x 44 C 1	x 1 C 0 x 12 C 1	= 5 x	44 x 1 x 12 =				2,640	DIV 3
5 C 3 x 44 C 2	x 1 C 1 x 12 C 0	= 10 x	946 x 1 x 1 =				9,460	
5 C 3 x 44 C 2	x 1 C 0 x 12 C 1	= 10 x	946 x 1 x 12 =				113,520	DIV 4
5 C 2 x 44 C 3	x 1 C 1 x 12 C 0	= 10 x	13,244 x 1 x 1 =				132,440	
5 C 2 x 44 C 3	x 1 C 0 x 12 C 1	= 10 x	13,244 x 1 x 12 =				1,589,280	DIV 5
5 C 1 x 44 C 4	x 1 C 1 x 12 C 0	= 5 x	135,751 x 1 x 1 =				678,755	
5 C 1 x 44 C 4	x 1 C 0 x 12 C 1	= 5 x	135,751 x 1 x 12 =				8,145,060	
5 C 0 x 44 C 5	x 1 C 1 x 12 C 0	= 1 x	1,086,008 x 1 x 1 =				1,086,008	DIV 6
5 C 0 x 44 C 5	x 1 C 0 x 12 C 1	= 1 x	1,086,008 x 1 x 12 =				13,032,096	

TOTAL POSSIBLE OUTCOMES: 24,789,492

$C_{(49,5)} \times C_{(13,1)} = 1,906,884 \times 13 = 24,789,492$

Prize structure for 5/49 + 1/13 as defined below:

CATEGORIAS DE PRÉMIOS							
	Prémios	1.º	2.º	3.º	4.º	5.º	Nº da Sorte
Acertos	Números	5	5	4	3	2	0
	Nº da sorte	1	0	0	0	0	1

6/40 (1 supplementary number) Plus 1/10

MACHINE A W	S		MACHINE B W			W	S		W			Result	
6 C 6	× 1 C 0	× 33 C 0	× 1 C 1	× 9 C 0	=	1 × 1 ×		1 × 1 × 1	=			1	DIV 1
6 C 6	× 1 C 0	× 33 C 0	× 1 C 0	× 9 C 1	=	1 × 1 ×		1 × 1 × 9	=			9	
6 C 5	× 1 C 1	× 33 C 0	× 1 C 1	× 9 C 0	=	6 × 1 ×		1 × 1 × 1	=			6	DIV 2
6 C 5	× 1 C 1	× 33 C 0	× 1 C 0	× 9 C 1	=	6 × 1 ×		1 × 1 × 9	=			54	
6 C 5	× 1 C 0	× 33 C 1	× 1 C 1	× 9 C 0	=	6 × 1 ×		33 × 1 × 1	=			198	DIV 3
6 C 5	× 1 C 0	× 33 C 1	× 1 C 0	× 9 C 1	=	6 × 1 ×		33 × 1 × 9	=			1,782	
6 C 4	× 1 C 1	× 33 C 1	× 1 C 1	× 9 C 0	=	15 × 1 ×		33 × 1 × 1	=			495	DIV 4
6 C 4	× 1 C 1	× 33 C 1	× 1 C 0	× 9 C 1	=	15 × 1 ×		33 × 1 × 9	=			4,455	
6 C 4	× 1 C 0	× 33 C 2	× 1 C 1	× 9 C 0	=	15 × 1 ×		528 × 1 × 1	=			7,920	DIV 5
6 C 4	× 1 C 0	× 33 C 2	× 1 C 0	× 9 C 1	=	15 × 1 ×		528 × 1 × 9	=			71,280	
6 C 3	× 1 C 1	× 33 C 2	× 1 C 1	× 9 C 0	=	20 × 1 ×		528 × 1 × 1	=			10,560	DIV 6
6 C 3	× 1 C 1	× 33 C 2	× 1 C 0	× 9 C 1	=	20 × 1 ×		528 × 1 × 9	=			95,040	
6 C 3	× 1 C 0	× 33 C 3	× 1 C 1	× 9 C 0	=	20 × 1 ×		5,456 × 1 × 1	=			109,120	DIV 7
6 C 3	× 1 C 0	× 33 C 3	1 C 0	× 9 C 1	=	20 × 1 ×		5,456 × 1 × 9	=			982,080	
6 C 2	× 1 C 1	× 33 C 3	× 1 C 1	× 9 C 0	=	15 × 1 ×		5,456 × 1 × 1	=			81,840	
6 C 2	× 1 C 1	× 33 C 3	1 C 0	× 9 C 1	=	15 × 1 ×		5,456 × 1 × 9	=			736,560	
6 C 2	× 1 C 0	× 33 C 4	× 1 C 1	× 9 C 0	=	15 × 1 ×		40,920 × 1 × 1	=			613,800	
6 C 2	× 1 C 0	× 33 C 4	1 C 0	× 9 C 1	=	15 × 1 ×		40,920 × 1 × 9	=			5,524,200	
6 C 1	× 1 C 1	× 33 C 4	× 1 C 1	× 9 C 0	=	6 × 1 ×		40,920 × 1 × 1	=			245,520	
6 C 1	× 1 C 1	× 33 C 4	1 C 0	× 9 C 1	=	6 × 1 ×		40,920 × 1 × 9	=			2,209,680	
6 C 1	× 1 C 0	× 33 C 5	× 1 C 1	× 9 C 0	=	6 × 1 ×		237,336 × 1 × 1	=			1,424,016	
6 C 1	× 1 C 0	× 33 C 5	1 C 0	× 9 C 1	=	6 × 1 ×		237,336 × 1 × 9	=			12,816,144	
6 C 0	× 1 C 1	× 33 C 5	× 1 C 1	× 9 C 0	=	1 × 1 ×		237,336 × 1 × 1	=			237,336	
6 C 0	× 1 C 1	× 33 C 5	1 C 0	× 9 C 1	=	1 × 1 ×		237,336 × 1 × 9	=			2,136,024	
6 C 0	× 1 C 0	× 33 C 6	× 1 C 1	× 9 C 0	=	1 × 1 ×		1,107,568 × 1 × 1	=			1,107,568	
6 C 0	× 1 C 0	× 33 C 6	1 C 0	× 9 C 1	=	1 × 1 ×		1,107,568 × 1 × 9	=			9,968,112	

TOTAL POSSIBLE OUTCOMES =	38,383,800
C(40,6) x C(10,1) =	38,383,800

6/49 (1 supplementary number) Plus 1/10

MACHINE A W	MACHINE A S	(42)	MACHINE B W	MACHINE B	Calculation	Outcomes	Outcomes	Probabilities 1 in	Prize
$_6C_6$	$_1C_0$	$_{42}C_0$	$_1C_1$	$_9C_0$	1 × 1 × 1 × 1 × 1	1	1	139,838,160	1st PRIZE
$_6C_6$	$_1C_0$	$_{42}C_0$	$_1C_0$	$_9C_1$	1 × 1 × 1 × 1 × 9	9	9	15,537,573	DIV 1
$_6C_5$	$_1C_1$	$_{42}C_0$	$_1C_1$	$_9C_0$	6 × 1 × 1 × 1 × 1	6	60	2,330,636	DIV 2
$_6C_5$	$_1C_1$	$_{42}C_0$	$_1C_0$	$_9C_1$	6 × 1 × 1 × 1 × 9	54			
$_6C_5$	$_1C_0$	$_{42}C_1$	$_1C_1$	$_9C_0$	6 × 1 × 42 × 1 × 1	252	2,520	55,491	DIV 3
$_6C_5$	$_1C_0$	$_{42}C_1$	$_1C_0$	$_9C_1$	6 × 1 × 42 × 1 × 9	2,268			
$_6C_4$	$_1C_1$	$_{42}C_1$	$_1C_1$	$_9C_0$	15 × 1 × 42 × 1 × 1	630	135,450	1,032	DIV 4
$_6C_4$	$_1C_1$	$_{42}C_1$	$_1C_0$	$_9C_1$	15 × 1 × 42 × 1 × 9	5,670			
$_6C_4$	$_1C_0$	$_{42}C_2$	$_1C_1$	$_9C_0$	15 × 1 × 861 × 1 × 1	12,915			
$_6C_4$	$_1C_0$	$_{42}C_2$	$_1C_0$	$_9C_1$	15 × 1 × 861 × 1 × 9	116,235			
$_6C_3$	$_1C_1$	$_{42}C_2$	$_1C_1$	$_9C_0$	20 × 1 × 861 × 1 × 1	17,220	2,468,200	57	DIV 5
$_6C_3$	$_1C_1$	$_{42}C_2$	$_1C_0$	$_9C_1$	20 × 1 × 861 × 1 × 9	154,980			
$_6C_3$	$_1C_0$	$_{42}C_3$	$_1C_1$	$_9C_0$	20 × 1 × 11,480 × 1 × 1	229,600			
$_6C_3$	$_1C_0$	$_{42}C_3$	$_1C_0$	$_9C_1$	20 × 1 × 11,480 × 1 × 9	2,066,400			
$_6C_2$	$_1C_1$	$_{42}C_3$	$_1C_1$	$_9C_0$	15 × 1 × 11,480 × 1 × 1	172,200			
$_6C_2$	$_1C_1$	$_{42}C_3$	$_1C_0$	$_9C_1$	15 × 1 × 11,480 × 1 × 9	1,549,800			
$_6C_2$	$_1C_0$	$_{42}C_4$	$_1C_1$	$_9C_0$	15 × 1 × 111,930 × 1 × 1	1,678,950			
$_6C_2$	$_1C_0$	$_{42}C_4$	$_1C_0$	$_9C_1$	15 × 1 × 111,930 × 1 × 9	15,110,550			
$_6C_1$	$_1C_1$	$_{42}C_4$	$_1C_1$	$_9C_0$	6 × 1 × 111,930 × 1 × 1	671,580			
$_6C_1$	$_1C_1$	$_{42}C_4$	$_1C_0$	$_9C_1$	6 × 1 × 111,930 × 1 × 9	6,044,220			
$_6C_1$	$_1C_0$	$_{42}C_5$	$_1C_1$	$_9C_0$	6 × 1 × 850,668 × 1 × 1	5,104,008			
$_6C_1$	$_1C_0$	$_{42}C_5$	$_1C_0$	$_9C_1$	6 × 1 × 850,668 × 1 × 9	45,936,072			
$_6C_0$	$_1C_1$	$_{42}C_5$	$_1C_1$	$_9C_0$	1 × 1 × 850,668 × 1 × 1	850,668			
$_6C_0$	$_1C_1$	$_{42}C_5$	$_1C_0$	$_9C_1$	1 × 1 × 850,668 × 1 × 9	7,656,012			
$_6C_0$	$_1C_0$	$_{42}C_6$	$_1C_1$	$_9C_0$	1 × 1 × 5,245,786 × 1 × 1	5,245,786			
$_6C_0$	$_1C_0$	$_{42}C_6$	$_1C_0$	$_9C_1$	1 × 1 × 5,245,786 × 1 × 9	47,212,074			

TOTAL POSSIBLE OUTCOMES = 139,838,160
C(49,6) x C(10,1) = 139,838,160

6/49 Plus 1/10

MACHINE A W	(43)	MACHINE B W	(9)	Calculation	Outcomes	Probabilities 1 in	Div
$_6C_6$	$_{43}C_0$	$_1C_1$	$_9C_0$	1 × 1 × 1 × 1 × 1	1	139,838,160	DIV 1
$_6C_6$	$_{43}C_0$	$_1C_0$	$_9C_1$	1 × 1 × 1 × 9	9	15,537,573	DIV 2
$_6C_5$	$_{43}C_1$	$_1C_1$	$_9C_0$	6 × 43 × 1 × 1	258	542,008	DIV 3
$_6C_5$	$_{43}C_1$	$_1C_0$	$_9C_1$	6 × 43 × 1 × 9	2,322	60,223	DIV 4
$_6C_4$	$_{43}C_2$	$_1C_1$	$_9C_0$	15 × 903 × 1 × 1	13,545	10,324	DIV 5
$_6C_4$	$_{43}C_2$	$_1C_0$	$_9C_1$	15 × 903 × 1 × 9	121,905	1,147	DIV 6
$_6C_3$	$_{43}C_3$	$_1C_1$	$_9C_0$	20 × 12341 × 1 × 1	246,820	567	DIV 7
$_6C_3$	$_{43}C_3$	$_1C_0$	$_9C_1$	20 × 12341 × 1 × 9	2,221,380	63	DIV 8
$_6C_2$	$_{43}C_4$	$_1C_1$	$_9C_0$	15 × 123,410 × 1 × 1	1,851,150	76	DIV 9
$_6C_2$	$_{43}C_4$	$_1C_0$	$_9C_1$	15 × 123,410 × 1 × 9	16,660,350		
$_6C_1$	$_{43}C_5$	$_1C_1$	$_9C_0$	6 × 962,598 × 1 × 1	5,775,588		
$_6C_1$	$_{43}C_5$	$_1C_0$	$_9C_1$	6 × 962,598 × 1 × 9	51,980,292		
$_6C_0$	$_{43}C_6$	$_1C_1$	$_9C_0$	1 × 6,096,454 × 1 × 1	6,096,454		
$_6C_0$	$_{43}C_6$	$_1C_0$	$_9C_1$	1 × 6,096,454 × 1 × 9	54,868,086		

TOTAL POSSIBLE OUTCOMES = 139,838,160
C(49,6) x 9(10,1) = 139,838,160

BIBLIOGRAPHY

Ashton, J 1893, *A history of English lotteries*, The Leadenhall Press, London.

Bender, E 1938, *Tickets to fortune. The story of sweepstakes, lotteries, and contests*, Modern Age Books, New York.

Bennett, D J 1998, *Randomness*, Harvard University Press, Cambridge, Massachusetts.

Blanche, E 1949, *You cannot win. Facts and fallacies about gambling*, Public Affairs Press, New York.

Catlin, D 2003, *The lottery book. The truth behind the numbers*, Supplementary Books, Chicago.

Charlton, P 1987, *Two flies up a wall. The Australian passion for gambling*, Methuen Haynes, Sydney.

Cheung, Y H 1996, *The unfavourable lotteries that people play and why they play them*, Edith Cowan University, Joondalup, Western Australia.

Crabb, T 1994, *Gambling to win in Australia*, Harper Collins Publishers, Sydney.

David, F N 1962, *Games, gods, and gambling. A history of probability and statistical ideas*, Dover Publications, New York

Eisler, H 1990, *Gambling into the nineties. A guide to better odds*, Kangaroo Press, Sydney.

Eisler, H 1992, *Improve your chances on poker machines*, Kangaroo Press, Sydney.

Jones, J P 1973, *Gambling yesterday and today. A complete history*, David & Charles, Newton.

Karcher, A J 1989, *Lotteries*, Transaction Publishers, New Brunswick, New Jersey.

Kreyszig, E 1970, *Introductory mathematical statistics. Principles and methods*, John Wiley & Sons, Singapore.

Levinson, H C 2001, *Chance, luck, and statistics. The science of chance*, Dover Publications, New York.

Livio, M 2009, *Is God a mathematician?* Simon & Schuster, New York.

McColl, J H 1995, *Probability*, Edward Arnold, London.

Moroney, M J 1990, *Facts from figures*, 2[nd]edn, Penguin Books, London.

Ore, O 1953, *Cardano: the gambling scholar*, Princeton University Press, Princeton, New Jersey.

Orkin, M 1991, *Can you win?* W. H. Freeman and Company, New York.

Paulos, J A 1990, *Innumeracy. Mathematical illiteracy and its consequences*, Vintage Books, New York.

Scarne, J 1986, *Scarne's new complete guide to gambling*, Fireside, London.

Solonsch, M, 1991, *The complete guide to Australian gambling*, Wrightbooks, Brighton, Victoria.

Sullivan, G 1972, *By chance a winner. The history of lotteries*, Dodd, Mead & Company, New York.

Todhunter, I 1865, *A History of the mathematical theory of probability from the time of Pascal to that of Laplace*, MacMillan and Company, London.

Weaver, W 1963, *Lady Luck. The theory of probability*, Dover Publications, New York.

Whiting, L 1997, *Analytical model of a combinatorial number lottery*, Rutledge Books, Bethel.

Printed in the USA
CPSIA information can be obtained
at www.ICGtesting.com
LVHW051051170823
755275LV00007B/572